CLIMATE CHANGE

CLIMATE CHANGE

Is It Really Caused by Carbon Dioxide?

DR. SAM O. OTUKOL

Ordering Information:

For orders and inquiries, please contact:
1-888-404-1388
www.goldtouchpress.com
book.orders@goldtouchpress.com

Printed in the United States of America

This book is dedicated to my mom Majori Okirya and dad Erenesti Okirya. For the two of them, their life-long commitment was to education of their children. My dad begged and borrowed from friends and sympathizers to make sure we had good education. They were both far sighted and ensure that both the sons and daughters had fair and equal access to education.

My mom worked ten long hours each day without rest. Many of those hours were under a hot sun, to ensure we had enough food to eat. And indeed we never lacked anything. Most of my siblings went to school and completed successfully, but a few struggled. Overall our parents were satisfied with the results of their efforts. That I was able to focus and finish this book, is mainly due to their inspiration. They gave me the spirit to never give up and to stand firmly for what I believed in.

Unfortunately, both passed on before they enjoyed the benefit of their hard work. May they both Rest In Peace. AMEN.

CONTENTS

PREFACE

My inspiration to write this book comes from my advanced studies in geography, chemistry, and biology, personal experiences with drought, and the twenty-two years I spent working for BC Public Service. I was born in Butebo in Uganda and completed a bachelor's of science degree in forestry. I moved to Canada in 1975 to complete a master's in science in forest biometrics at the University of New Brunswick. From there, I moved to the University of Toronto to complete a PhD. My work has involved designing statistically robust methods for collecting natural resource data, writing data analysis procedures, and advising decision makers on how to interpret the statistics from the collected data. I learned that decision makers quite often make critical decisions based on what information is available, even when it might have known defects, and that the key skill is in managing risk and uncertainty of using the data. Such is the case with climate change data. There is much more to climate change than increasing carbon dioxide. This book explores what else is behind it.

LIST OF FIGURES

LIST OF TABLES

INTRODUCTION

At the beginning of this third millennium, AD, the world is in crisis. As the leading authority on climate change, the Intergovernmental Panel on Climate Change (IPCC) has concluded that there is clear indication that human activities are causing real changes in global weather. They state their conclusions bluntly: "Warming of the climate system is unequivocal, as is now evident from observation of increases in global average air and ocean temperatures, widespread melting of snow and ice, and rising global average sea level."

The IPCC lays the blame for global warming squarely at the feet of human activity (i.e., industrial production), generating what is now known as greenhouse gases (*Climate Change: The Physical Science Basis*, IPCC, 2007). The most important gas has been identified as carbon dioxide; its levels have been observed to be increasing rapidly over the past one hundred years. The other gases include methane and nitrous oxide. The position of the IPCC and other proponents of the greenhouse effect theory will be discussed in more detail in the section on current theories.

In spite of the IPCC's reports, testimonies from many scientists, and the regular newspaper reports of changes occurring around the globe, some detractors are convinced that the claims regarding climate change are not real. In most cases, the nonbelievers fall into two categories. On the one hand, die-hard politicians want to defend our current way of life via an ideology that asserts that our activities, particularly commercial activities, should not be interfered with in any way. On the other hand, another group believes that all these changes in weather are temporary, caused by minor changes in the sun's radiation, and that global climate will go back to normal in a few years. This book presents a different theory on what might be happening. It is quite likely that climate change is a continuous process that has been going on in cycles and may have happened earlier in history when there were no industrialized civilizations. The question is, what is the common factor between those earlier occurrences of climate change and the current occurrence?

Climate change scientists have not been forthcoming in explaining the mechanism by which carbon dioxide and the other greenhouse gases hold heat.

The other factor to consider is that carbon dioxide molecules are known to be quite stable at atmospheric temperature. They do not change form or have an affinity to combine with other molecules at temperatures below 200 degrees Celsius. In view of this, it is difficult to rationalize the claim that this component of the atmosphere has great influence on global temperature. It is, however, fair to state that where water vapor is generated, so is carbon dioxide (CO_2). In that sense, CO_2 serves as a good indicator of how much water vapor is being generated.

The absorption of ultraviolet light by carbon dioxide may be an issue; however, the relative size of the contribution of carbon dioxide to the atmospheric gases limits its power in causing dramatic changes to climate.

This book holds the view that this is a "who done it" or whodunit detective story; it is possible that the wrong suspect may be thrown in jail based on circumstantial evidence but that the real culprit might be a gas that constantly hangs out with the suspect.

This book will also show that the culprit has been known to form bonds with carbon dioxide and that breaking those bonds creates a considerable amount of heat. The heat created during the breaking of the bond changes the behavior of the culprit dramatically, and this change in behavior is at the heart of our climate change problem.

This book is based on known principles of physics. The only problem is that the culprit is such a nice guy that it is difficult to believe that he/she has a dark side. Hopefully this book will stimulate worldwide interest among readers and get them to take a closer look at where the true culprit has been, and why ignoring him/her might hurt us badly.

The title suggests a hint of humor in the subject of global warming. It is not, however, possible to be amusing on this subject. The potential outcomes of the phenomenon include the possibility of mass relocation of populations from areas that will be hardest hit, mass starvation, homelessness, and misery on a scale we have never witnessed before. So it is not a matter that should be taken lightly.

The IPCC (www.ipcc.com) is by far the ultimate repository of the best information on climate change and global warming. The documentation from the site suffers from only one disadvantages— the use of technical language. Their most basic document, the 2007 "Summary for Policymakers," contains jargon such as "anthropogenic warming," "radioactive forcing," and so on. The use of such technical terms is avoided as much as possible in this treatise, as is the use of equations.

WHAT IS THIS BOOK ABOUT?

This book is about climate change (or global warming), as has been presented by the IPCC, newspapers, and other sources. Climate change includes increasing temperatures, fluctuations in rainfall and snowfall, changes in ecosystems, and changes in the composition of gases we inhale as they relate to change in global temperatures.

This book does not include discussion on air pollution, which is a separate issue altogether. Air pollution is the injection of particulate matter and gases that are not naturally expected into the atmosphere. In extreme circumstances, air pollution results in a condition called smog, whereby the polluting particles create a hazy air mix that blocks out sunlight. In these extreme situations, the hazy conditions have the opposite effect of global warming in that the haze blocks more than 80 percent of the sunlight from reaching the ground, without which the air does not warm up after sunrise. When sunlight rays reach the ground, they are transformed to create radiation, which we feel as heat.

Cities such as Los Angeles, Toronto, London, and others in Europe were at one time famous for the hazy air pollution conditions. More recent developments in automobile manufacturing technology and introduction of strict standards have virtually eliminated smog in most of these cities. There are still a few exceptions, such as Beijing, where smog conditions persist.

Combining discussion of global warming with discussion of air pollution tends to confuse issues and complicates efforts to mitigate global warming, because reducing air pollution does not necessarily translate into reduction in temperatures. Secondly, the emphasis on carbon release and sequestration seems to be more relevant to air pollution than it is to global warming. For instance, many forms of internal combustion engines release large quantities of soot into the air, causing pollution, but the soot does not result in increasing atmospheric air temperature. It is tempting to try to kill two birds with one stone, but it is risky because the best you might achieve is to get one bird and miss the other, and then have to deal with the negative consequences of not covering all bases.

Well, now we know there is something happening with climate change on a worldwide basis and that it is desirable for all 7 billion people to know what is happening and what can be done to return the climate to what was there before human activity upset the balance. But in order for anyone to do anything meaningful about it, we need to improve our knowledge of what is upsetting the climate.

At this point, climate, like a patient in a doctor's office, is believed to be suffering from imbalance in releasing and sequestering carbon dioxide, among other things. Producing too much carbon dioxide and not sequestering enough causes the imbalance.

It is suggested in this book that pumping too much water vapor into the atmosphere causes the imbalance in climate. As vapor is pumped into the atmosphere so is the heat of transformation that causes liquid water to change to gas.

The difference between the two opinions is significant because the prescriptions required to address the two causes are different. Reducing carbon emissions addresses the carbon release problem, but it does not necessarily address excess vapor production resulting from irrigation and the water warming effects of nuclear-power-generating plants.

If a patient goes to a doctor and is misdiagnosed, the consequences can be quite negative, and in some cases the error can be fatal. For this reason, it is appropriate to spend some time to make sure the diagnosis is right and provide the right prescription if possible.

EVIDENCE OF GLOBAL WARMING

Only a person who has been away from our planet for some time would claim that global warming is not happening. The evidence shows abundantly on all the continents of the world. Several aspects of climate have been affected. For instance, in the temperate zones, the summers are warmer and more humid. Droughts are more frequent, particularly in Africa. Rainfall and snowfall are more intense and, in some cases, quite destructive. Snow (or ice) cover on mountaintops is diminishing compared to what was there in 1950. There is scientific evidence showing that ice in the polar zones is melting at a faster rate.

A general continent-by-continent evaluation follows.

Africa

This continent is the only one that squarely straddles the equator and equally spans the north and south Tropics of Cancer and Capricorn, respectively. From the climate perspective, its orientation could be figuratively compared to a rider straddling a wild horse. The first European visitors to Africa described it as the Dark Continent. This rushed judgment has hampered African economic progress, yet it is based on ignorance, lack of education of African ecology and a hatred of an enigmatic people. In spite of harsh conditions it is difficult to understand why people in Africa still easily express happiness.

The African continent receives more solar radiation than any other on earth, and based on this knowledge, it is best described as the continent of light. By virtue of its location, the continent amplifies heat generated from solar radiation. This role of the continent should be studied because the results of such a study might be useful in mitigating the effects of climate change. To the north, the Sahara Desert occupies one-third of the continent. In the south, the Kalahari and Namib Deserts spread across close to one-tenth of the continent.

Figure 1. Map of Africa (source: www.pixaby.com)

Over the past two decades, the continent has been plagued by increasingly severe droughts. The Sahara is slowly creeping south into the Sahel zone. Between 2002 and 2006, several major droughts affected most of the African countries. The countries that were particularly hard hit included Eritrea, Ethiopia, Somalia,

Sudan, Namibia, Botswana, and Angola. For countries like Kenya and Uganda, only vulnerable regions were affected.

During the last spate of droughts, the level of Lake Victoria fell by close to five meters, stranding municipal water intake pipes. This left many municipalities, including Kampala and Jinja, with limited or no water supply. Lake Victoria is significant because it is the source of the River Nile. Significant drop in its level has a ripple effect on several countries along the River Nile basin.

In Namibia and Botswana, there was no significant amount of rainfall for two years.

Figure 2. Bush elephants in Botswana

Figure 3. Lake Victoria

As some countries suffered major water supply problems, record rainfall was recorded in Mozambique. In June 2006, an international effort was mounted to rescue people stranded by major flooding of the Zambezi River. Ethiopia, which experienced many droughts, suffered another major one during the 2002–2006 period.

Several major lakes on the African continent are drying up. Lake Chad is the most vulnerable, as it is so close to the Sahara Desert.

Lake Nakuru in Kenya is surprisingly one of the few lakes in Africa that are rising. This lake is a significant habitat for flamingos and other rare tropical birds. The shallowness of Lake Nakuru creates an ideal environment for plankton and other organisms that form the food supply for the flamingos. Unfortunately, the planktons in question do not thrive in deep waters. As such, the increased water level has a negative impact on bird and perhaps other wildlife habitats.

The construction of the Aswan Dam created Lake Nasser. This lake sits in an area where rainfall is rare, so the rate of its evaporation is perhaps higher than those of African lakes in cooler areas. Besides this, Egypt has tapped into the lake and created smaller lakes in the Sahara Desert to supply water to agricultural projects. The evaporation of water from Lake Nasser and the other lakes created in the Sahara could be significant factors in moderating climate.

Figure 4. Sand left behind by retreating Lake Chad

Figure 5. The snowcap on Mount Kenya

Figure 6. Mount Kilimanjaro

North America, Iceland, and Greenland

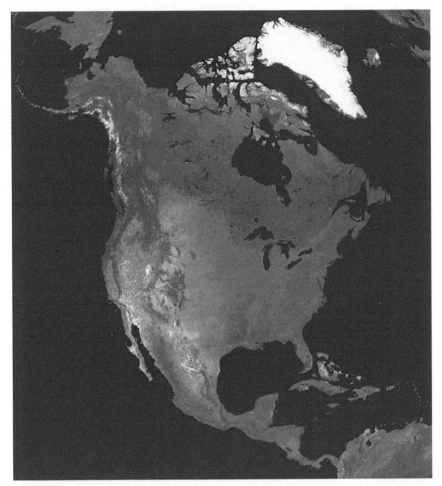

Figure 7. Continent of North America, Iceland, and Greenland

The imaginary Tropic of Cancer line in North America cuts an interesting path. It starts in the west at Cabo San Lucas, traverses briefly over the pacific to get to Mazatlan on mainland Mexico, brushes by Catorce, and exits the mainland just north of Tampico, from where it glides over the Gulf of Mexico, shooting past Cancun, and then heads into the Caribbean Sea, skipping over Cuba. This trajectory over the North American continent is largely over water, and it is probably the main reason why desertification is not as severe there as it is in Northern Africa along the same latitude. However, it may be that states in the proximity of the latitude in North America should be aware of the risk to desertification posed by the proximity to the line. The Caribbean need not worry about this since it is located on the eastern side of the continent. Their concern should be in regards to tornadoes, hurricanes, flooding, and mudslides.

North America is home to the world's most prosperous economy—the United States (USA). Unfortunately, it is also a major center for global-warming skepticism. This is unfortunate, because as most people know, industrialization is a key factor in creating warmer weather. As a leading industrialized country, the United States is a leading contributor to the problem.

The reluctance of the US government to address global warming issues haunts the world.

Already, states in the southern part of the country are facing major water supply problems. In 2007, parts of the state of Georgia experienced a reduction in their water supply. The water level of Lake Manning, a major source of drinking water for many municipalities, dropped six meters (eighteen feet). The water table levels of several other freshwater lakes, such as Lake Mead, are dropping—sometimes quite dramatically.

All the states that are located along the southern border of the United States with Mexico or the Gulf of Mexico (including California, Arizona, New Mexico, and Texas) are facing drier climates. These states are at risk of running out of water at some point, if the weather trends continue the way they are. The drier weather has caused more frequent forest fire occurrences. Some of the fires in California in 2006 and 2007 were spectacular in their documented level of destruction. Some of the fires have even resulted in loss of life.

Figure 8. The Colorado River

Other than the Great Lakes, the United States does not have very large lakes that could supply the vulnerable states with a long-term water supply. The two major rivers in the country—the Colorado and the Mississippi—are vulnerable to the vagaries of snowfall in the Rockies and the Appalachian Mountains. Currently, it is known that snowfall in these mountain ranges is declining. The Colorado, in particular, is facing the possibility of drying up completely within the next century. The consequences for such an occurrence would be dire. For instance, California would cease to be the breadbasket of North America, and the price of their agricultural produce would skyrocket.

Surprisingly, the dry conditions in the southwestern and southeastern parts of the United States might be countered by excessive rainfalls that could become the order of the day in the era of global warming. According to estimates, a greater part of the Midwestern states will actually experience higher-than-average rainfall over the next fifty years. Some parts of the Midwestern states might become ephemeral lakes, flooding for several months and then remaining bone-dry for short spells of time. The 2008 floods in Iowa should be expected to occur on an annual basis. The hurricanes that have plagued the lower Mississippi will become more severe and more regular, occurring more frequently rather than just once every decade.

Figure 9. The Great Salt Lake in Utah

Figure 10. Lake Mead

In the meantime, in Alaska, the breakup of glaciers is occurring at unprecedented rates. Some areas where the permafrost condition was a valuable support for winter transportation will be altered permanently.

Elsewhere in North America, the warmer weather has encouraged forest-pest infestations. In British Columbia, Canada, the mountain pine beetle infestation has decimated one-third of the potential timber-supply source. The possibility of other outbreaks of similar pests has been assessed as quite high.

At the North Pole, there is an unprecedented breakup of ice that used to cover the Northwest Passage even during the summer months. If the warming trend continues, the passage will become accessible to commercial shipping. This may sound like good news, but the downside is that the habitat for polar bears may become endangered.

Figure 11. Retreating glaciers

Since much of Mexico has always been either desert or semidesert, the drier areas in the country will remain that way.

Iceland and Greenland, the two countries with permanent ice cover, are experiencing rapid ice melting. The loss of ice in these subcontinents may result in major changes in sea level for some coastal areas around the world. In some low-lying places, such as Bangladesh, major flooding may result.

South America

South America consists of some of the few countries on track to join developed countries in terms of economic prosperity. Particularly, Brazil is not only industrializing at an alarming rate but is also conducting an aggressive assault on forested areas of the Amazon.

In a world where pursuing national interests supersedes the need to protect the environment, Brazil cannot be blamed for trying to raise the standards of living for its people.

Figure 12. Map of South America

Figure 13. Atacama Desert in South America

The Tropic of Capricorn enters the South American continent near the city of Antofagasta on the west side of the Atacama Desert, runs over the Andes Mountains, crosses northern Argentina, slices Paraguay in half, runs through southern Brazil, and exits the continent near Sao Paulo. The significance of this description is that there is a very dry desert at the west extreme of the twenty-third-degree latitude. In Wikipedia, the Atacama Desert is described as the driest nonpolar desert in the world. This desert is very dry because the desertification did not allow development of river systems on the western side of the Andes in that area, and any lakes that may have existed prior to desertification became salt flats.

Here in South America, as is the case in Africa and Australia, the Tropic of Capricorn is associated with desertification.

The saving grace for South America is the presence of the Andes Mountains. These mountains are high enough, and the land west of them is relatively flat, allowing moisture from the Atlantic and Amazon basin to prevent the spread of the desert farther east.

As a consequence, the main impact of climate change on the South American continent will be in regard to torrential rains, mudslides, and hurricanes.

Europe

Of all the countries in Europe, Britain has always boasted of having moderate weather, apparently caused by the fact that it is surrounded by ocean. The cool ocean moderates the summers, so they are not too warm. Conversely, warm currents around the country make the winters relatively warmer than elsewhere in Europe. The moderate weather is no longer the norm in Britain. In 2009, the summer was too hot. In the winter, there was record snowfall and record-breaking cold temperatures.

Elsewhere in Europe, particularly in France, Spain, and Portugal, the elements have not been very kind. Not only have there been heat waves, but record-breaking rainfall has caused floods and washed away bridges. In Spain, farm areas in the south have experienced prolonged droughts during the crop-growing season.

In cities such as Venice, residents fear the threat of rising sea levels.

Russia

Russia is a large industrialized country with many vapor-generating industries, including nuclear energy and agricultural industries.

The most widely published incident resulting from global warming pertains to the steady disappearance of the Aral Sea. In this incident, water evaporated from most of the Aral Sea, leaving dry land behind. Close to 90 percent of the water disappeared into space. This incident is instructive in two ways. First, solar radiation is quite powerful and can cause considerable changes. Second, the atmosphere can hold large quantities of water for some time. The net effect of the Aral Sea water evaporation process was that the water was airlifted and dumped somewhere else, either in Europe or Asia.

The winters in Russia have also been unusual. For instance, the 2008 and 2009 snowfalls in Moscow were quite heavy, occasionally severe.

Figure 14. Aral Sea in Russia

Asia

Asia is well known for its monsoon winds. These regular winds have been so influential in history that at one point they were the key lifeline for trade between Europe and the Far East. Monsoons can bring good things, but occasionally they also bring catastrophe. Over the years, the regularity of the monsoons has not changed much, but the intensity of monsoon storms seems to have increased. Besides record rainfall, the monsoons now spawn typhoons more frequently than they did previously.

In 2008, monsoons caused typhoons in Myanmar (Burma). Vast areas of land were flooded, and many people and livestock were lost. In addition, considerable farmland and crops were destroyed.

In 2009, a vast area, including Manila, the capital city of the Philippines, received record rainfall that persisted for days. Thousands of people were left homeless, and there was considerable damage to infrastructure and property.

Over the past ten years, Bangladesh, a country located at the delta of the River Ganges, has experienced several catastrophic floods. This country suffers from two unfortunate circumstances: the first is its location at the confluence of two major rivers that drain from mountainous neighboring countries. The deltas of these rivers flood during the rainy seasons. The second is its location at sea level, adjacent to the Indian Ocean. Increases in rainfall have serious consequences for this nation.

In 2008 and again in 2009, China experienced record-breaking snowfalls and rainfalls. Heavy rainfall has damaged infrastructure, and the snowfalls paralyzed transportation systems and caused fuel shortages.

In 2009, Iran experienced an unusually cold winter; the ripple effect caused the country, which is a major supplier of fuel, to run out of fuel.

Middle East

Throughout history, the Middle East has recorded stories of drought and desertification. This cycle seems to continually progress; in fact, it seems to be accelerating. For instance, the water level of the Dead Sea has been dropping steadily over the past two decades. It is estimated that the seawater level has dropped by forty meters since 1950, and the level continues dropping at an alarming rate of one meter per year. Now hotels that were located at beachside by the sea have to ferry their clients to the sea by bus.

Figure 15. The Dead Sea

The rainfall cycles that many peasant farmers had become accustomed to and planned for do not exist anymore. Rainfall comes unexpectedly and quite often in torrents that sweep away homes and ruin crops. Water scarcity in the Middle East in countries such as Saudi Arabia, Jordan, Iraq, Yemen, and Israel has been part of their history. However, climate change may exacerbate matters. Water security issues may increase tensions in an area that has already seen much grief.

Australia

The Tropic of Capricorn slices through Australia, almost cutting the continent into two equal parts—one half north and the other half south of the twenty-three-degree latitude. The latitude enters Australia at the extreme eastern part of the continent, passes just south of Alice Springs, and exits just north of Brisbane. It is the second continent to Africa with such large land area in the sensitive tropical zones. As indicated before, it seems that these zones are susceptible to desertification. Given that most of the western part of Australia is desert, it seems that this assertion is confirmed.

Australia is one of the countries where the extreme impacts of climate change have been occurring over the past two decades. Not only have they had severe droughts followed by epic fires, they have also experienced typhoons, heavy rains, and severe flooding. Lying along the Tropic of Capricorn should explain the severe droughts. Similar drought conditions have been also happening in Africa, the continent with the largest area straddling the Tropic of Cancer and the Tropic of Capricorn.

It seems reasonable to speculate that these two continents should brace for worse droughts to come. In the worst-case scenario, some populations on these two continents may have to relocate to places where life is sustainable.

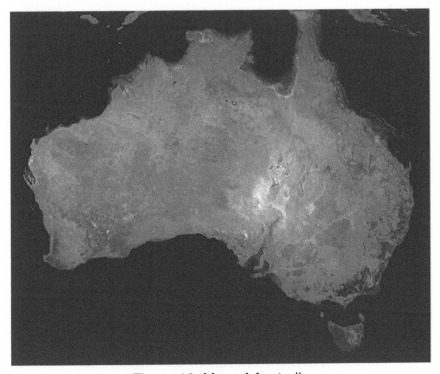

Figure 16. Map of Australia

Figure 17. Deserts in Australia

GREENHOUSE GASES

The current and most popular theory on global warming suggests that a number of gases in the atmosphere are trapping radiation, being reflected back into space, and preventing the normal cooling that traditionally occurs over the course of a day. Over several centuries, it seemed that global temperatures had stabilized. Minor annual fluctuations occurred, but there was no discernible trend toward temperature increase or decrease.

The earth, according to this theory, is normally warmed by solar light rays that arrive in bursts of shortwave energy radiation. The light is absorbed by rocks, trees, grass, water, and soil. The objects that absorb the light convert the shortwave radiation into long-wave bursts of energy—heat. Normally, the heat radiates back into space. Now it is claimed that an increase in some obnoxious gases (such as carbon dioxide, methane, water vapor, and nitrous oxide) is interfering with the heat's radiation back into space, causing global temperatures to increase.

The absorption of light by grasses and trees is a desirable process, reducing the amount of light that is converted to heat. In addition, the trees and grasses absorb carbon dioxide, which (according to the theory) is a key culprit in trapping heat. The plants store (or sequester) the light energy in the form of fuel or food so that it can be used later when needed.

The conclusion that carbon dioxide, water vapor, methane, nitrous oxide, and other gases are trapping heat was reached after various scientists, including several working for the IPCC, discovered that the levels of these gases were increasing along with global annual temperatures. It seems that there is a strong correlation between the increasing global temperature and the increasing levels of the named gases.

On average, the increase in global temperature seems to be fairly modest. Over the past ten years, the average global annual temperature has gone up by approximately one degree Celsius. A one- degree (Celsius) increase in annual global temperature seems to be quite small, but that is because some areas on the globe may be experiencing lower than normal annual temperatures while others are experiencing higher than normal temperatures. In fact, some local areas may be experiencing more dramatic decreases or increases in temperature for a given year.

In addition, such an increase in average global temperature, minor as it might seem, is expected to occur over a century or longer—not a decade. Thus, the increase is alarming to scientists. Beyond the scientists, most people have noticed changes in local weather fluctuations. The summers, in particular, have been unusually warm in the temperate areas of the globe, and many parts of the tropics have witnessed prolonged droughts.

The warmer weather has also been accompanied by heavier snow and rain precipitation. In some areas, the equivalent of an entire year's expected rainfall has occurred in nonstop rainfall over the duration of one to two weeks. In the temperate areas, snowfalls have been heavy and sometimes unpredictable.

It seems that the trend toward global warming has been happening for some time. If this is true, a reversal of the effects may not occur in the short term. It may take two to three decades for the mitigation efforts to have an impact. For that short-term period, it is likely that extreme weather changes will cause considerable damage to land and property. For this reason, everyone needs to pay attention, engage in discussions, and contribute positively by finding solutions to the problem.

The Role of Carbon Dioxide

The proponents of the current theory on global warming have identified carbon dioxide as the most significant contributor to global warming. If this gas were a person, he/she would have been in jail with no chance for parole for eternity.

In 1896, the famous scientist Svante Arrhenius (he won a Nobel Prize in chemistry in 1903) initiated pioneered work on tracking levels of carbon dioxide in the atmosphere. Since then, many scientists around the world now track levels of CO_2 as a matter of routine. Svante's theory was that carbon dioxide regulated the levels of water vapor in the atmosphere. The mechanism by which carbon dioxide regulates vapor is not well explained in literature.

At the core of most systems that forecast future global warming is a formula developed based on research by Svante Arrhenius. In 1896, he published a paper titled "On the Influence of Carbonic Acid in the Air upon the Temperature of the Ground" (*Philosophical Magazine and Journal of Science* 41, 237–276). In this publication, he explained his theory in detail. The essence of the theory is that carbon dioxide absorbs ultraviolet light from the sun and by so doing starts a cycle of warming of air called the greenhouse effect. More carbon dioxide in the air absorbs more ultraviolet light, and the heating increases.

In an equation, the theory is stated as follows (Svante 1896):

$$\Delta F = \alpha \ln (C/C_0)$$

where ΔF is a measure of the change in heating (also called radioactive forcing)

α is an expansion factor

C is existing carbon dioxide concentration

C_0 is carbon dioxide concentration before warming started

ln is a natural log transformation of what is calculated in the brackets Based on this equation, weather will get warmer as the levels of carbon dioxide increase above the baseline levels. The warming is complicated by vapor, which is generated as a by-product of the greenhouse effect.

In 1900, Knut Ångström published a paper titled "About the Importance of Water Vapour and Carbon Dioxide during the Absorption of the Earth's Atmosphere" (*Annalen der Physik* 308, no. 12: 720–732) that seemed to contradict Svante's work. A scientific discussion ensued, and some controversy still exists.

In spite of the possible contradictions raised by Knut Ångström in 1900, it is now generally accepted in scientific circles that this gas is the main villain of the crisis. It is also well known that the gas is toxic in high concentrations and will cause death to animal organisms if inhaled at a high concentration. However, the gas is indispensible in the processes that maintain plant and animal life.

With regard to plant life, carbon dioxide and water are essential for building the basic blocks of energy—starch, carbohydrates, and sugars. These are the compounds that power every living thing, be it plant or animal. In this role, carbon dioxide is vital; no fossil energy can be made without it.

Generally, carbon dioxide is heavier than the other gases in the atmosphere, including water vapor. If you had a jar full of air and you poured pure carbon dioxide into the jar, the carbon dioxide would sink to the bottom, and the other gases would float on top.

This heavier-than-air characteristic of carbon dioxide makes it unlikely for its accepted role in global warming. A gas that would be effective in insulating the earth would have to float slightly above the other gases in the atmosphere. In fact, water vapor tends to do just that, as can be determined from the position of clouds that float above ground.

Water vapor has been listed as one of the greenhouse gases, but many scientists have discounted it, saying that its increase in the atmosphere has been small (approximately 0.04 percent). In fact, many scientists view water vapor favorably, because as clouds, water vapor seems to absorb solar radiation and lower local temperatures.

As it hovers close to the ground, water vapor constitutes 50 to 90 percent of the air by volume, depending on temperature and local availability of vapor. The air's ability to carry vast quantities of water in vapor form could be perceived as one of the key mysteries to the existence of life on earth. As a vapor, water can be moved about more easily and can be taken to heights where, ordinarily, it would be impossible for it to reach.

It is also conceivable that water vapor in the atmosphere is the key force behind daily weather changes. To begin with, the transformation of liquid water to vapor results in considerable change in the volume of space occupied by water. The transformation is similar to the change that occurs when solid dynamite is transformed to gas, except that the water transformation is not as violent in open space. In closed space, transformation of liquid water to ice has so much force that it can break up big rocks.

For the transformation of liquid water to gas, considerable energy from the sun is required. In most cases, soil mixed with liquid water absorbs the sun's energy, which heats the water, and then in an instant—vapor is formed. In other cases, liquid water in lakes or oceans absorbs the sun's energy, and the transformation occurs all day directly from water bodies. This transformation of liquid water to vapor is pervasive and occurs every twenty-four hours at all temperatures, including temperatures below 0 degrees Celsius.

Since water vapor is only formed as a result of liquid water absorbing energy, it is fair to conclude that water vapor is a store of heat energy. If the water vapor is to be transformed back to liquid water, the absorbed heat must be transferred to a cooler entity. The heat transfer usually occurs quietly in the atmosphere when cooler water vapor molecules collide with a warmer body of air or object.

Clouds

Figure 18. Clouds near Pender Island, BC
(Photo by Dr. Sam Otukol)

In nature, water presents itself in many forms and states. In its solid form (ice), it is transparent but can occasionally appear milky. In its crystal form (snow), the water is solid, but the arrangement of the molecules is different from its solid ice form. In its liquid form, present in oceans, lakes, and rivers, water is most abundant; it covers 75 percent of the earth's surface.

In its gaseous form, which requires a transformation of liquid into gas, water displays unique characteristics. A cubic meter of water weighs 1,000 kilograms (kg). But when this liquid is transformed to gas, the one cubic meter expands to occupy anywhere from 1,000 to 1,600 cubic meters of space. So one cubic meter of air, containing a high concentration of water vapor, may weigh between 0.5 kilograms to 1 kilogram.

The transformation to gas makes water lighter and causes it to float in the air. In this form, water molecules behave more like hydrogen molecules. As a result, water molecules are able to rise and float over other gases, such as hydrogen, carbon dioxide, and oxygen.

The second characteristic of water vapor is that it can become visible under specific circumstances. These circumstances are a high concentration of the vapor in the air and cold temperatures.

Clouds are neither liquid nor solid. They are still a form of gas. If the vapor in a cloud turns into liquid water, rainfall occurs, because raindrops are heavier than the gases in the atmosphere. If clouds were solid, airplanes would not be able to fly through them. The shift from invisible vapor to visible cloud is yet another transformation. Thick clouds are opaque (i.e., light can barely pass through them).

The rearrangement of water vapor molecules to form clouds is triggered by a change in temperature. As vapor rises and temperature drops, a threshold temperature is reached as water molecules regroup themselves into cluster-creating clouds. The regrouping of molecules is yet another transformation.

Types of Clouds

There are generally five types of clouds, including mist and fog. Mist appears a few meters above ground and is generally lighter than fog. Fog forms at ground level and tends to be quite thick. In many places, particularly in Europe and the United States, thick fog has caused major traffic accidents because it reduces visibility considerably.

Figure 19. Fog on top of Mount Douglas
(Photo by Dr. Sam Otukol)

At higher altitudes, there are three major types of clouds, distinguished by their physical appearance and the altitude at which they appear. The most distinguishable clouds are called *cumulus*. They are puffy and have a definite three-dimensional profile. Generally, they have a flat bottom and a rounded top like the top of a mushroom.

By their appearance, it seems that they appear directly above a ground source of evaporated water. It seems that vapor shoots up from the ground, rising until it reaches an elevation where low temperature transforms invisible vapor to visible cloud. Cumulous clouds appear at an elevation of 500 meters to 3,000 meters above ground, and in most cases, they seem to be stationary or have very little lateral motion. But variations do occur, particularly when wind moves them. When they are in motion and encounter colder air, rainfall will occur.

The second type of high-elevation cloud is the *cirrus*. These appear at an elevation of 3,000 meters to 6,000 meters above ground. They usually appear stretched out and thin. They are, in fact, cumulous clouds that have been moved about by wind.

The third type of high-elevation cloud is the *cirrostratus*. These appear at elevations between 8,000 meters to 10,000 meters above ground. They are light clouds, generally gray in appearance. They occur at an elevation at which winds are considerably stronger; as a result, they may appear as patchy sheets in the sky.

There are many variations to cumulus, cirrus, and cirrostratus clouds described above. More importantly, they are the strongest evidence to show that there is a considerable amount of water in the atmosphere. They are the most significant means by which fresh water is distributed over land. All rivers and most lakes are sustained entirely by water vapor molecules moving freely from one place to another.

At very high altitudes on top of mountains, water vapor appears abundant even when gases such as oxygen and carbon dioxide are scarce. As a result, most mountains' tops are covered in snow or ice. From this observation, it appears that hot water vapor tends to behave more like hydrogen, and the addition of an oxygen atom to the two hydrogen atoms does not seem to affect its lightness.

It is well known that there is a lower concentration of oxygen at elevations above 4,000 meters above sea level. Oxygen is lighter than carbon dioxide, so the observation regarding oxygen would also apply to carbon dioxide. The observation of abundant water vapor at such high elevations is a unique characteristic that proves the dynamic and light nature of water vapor.

The transformation of water from one state to another requires either an input of heat energy or a removal of that heat energy. For instance, heat is required to convert ice, a solid, to liquid water. Furthermore, heat energy is required to convert liquid water to vapor—a gas.

The reversal of the process from vapor (gas) to solid requires the removal of the heat that caused the transformation to gas in the first place. In a sense, transformations of water to different forms are powerful dynamics in nature.

There are two key factors in nature driving the water transformations. On the input side is solar radiation. It is absorbed by water molecules and causes a transformation from ice to liquid water and from liquid to vapor. Once that energy is absorbed and transformation occurs, the energy is stored in the vapor molecules as latent energy.

If we take the disappearance of the Aral Sea as an example, we should look at some of the reasons why it disappeared. The top reason is the diversion of tributary rivers that supplied the sea with fresh water. The river water was diverted for irrigation and agricultural use. The second possible reason is the impact of clearing of trees from large tracts of land for conversion to industrialized agriculture. In addition, Russia is a big industrialized country whereby considerable heat is generated from automobiles, locomotives, processing plants, nuclear energy generation, and enhanced domestic boiling of water. The combination of these factors made the use of the sea unsustainable, resulting in excessive water evaporation.

It is appropriate to look a bit deeper into the evaporation process itself. The Aral Sea was a large water body, and it had large seagoing vessels plowing its waters. As such, the sea had a huge amount of water in it. If we assume that it contained 1.5 billion liters of water prior to the accelerated evaporation, and one billion liters were evaporated, we can calculate the amount of energy that was required to covert liquid water to vapor.

In physics, it is known that approximately 622 kilocalories of heat energy are required to convert one liter of liquid water to vapor. So in the case of the Aral Sea water reduction, 622 billion kilocalories of heat energy were required to convert the one billion liters of seawater to vapor. That amount of energy was stored in the vapor as latent energy, and given that vapor can travel in any of the three possible dimensions (i.e., upward, north, east, south, and west), that huge energy reservoir could have been transported to any part of the globe, where it would be expended when the vapor is converted back to liquid in a rainfall. In the event of the vapor being converted to snow, the amount of heat energy expended would be even larger than the 622 kilocalories per liter.

If you consider that large bodies of water such as the Dead Sea, Lake Mead, Lake Victoria, Lake Chad, and many others have had large net losses of water leading to drops in the levels of those water bodies, it is clear that the dynamics of heat transfer to the atmosphere have been stupendous over the past two decades. Liquid water requires heat energy to power the transformation from liquid to vapor. The absorbed heat energy is transferred to cooler vapor molecules when rain is formed from vapor.

In the case of the Aral Sea example, 622 billion kilocalories of energy were absorbed at the lake site to create vapor. But this vapor travelled thousands of kilometers from Aral, and as it was hovering at a place far away, in Alaska for example, it released the latent heat in order to be transformed to snow or rain. More heat would have been released if the transformation were from vapor to snow than would have been if the transformation were from vapor to rain.

Prior to industrialization, the energy used to create evaporation balanced well with the energy released during condensation to create rain. However, as more vapor is added from new industrial processes, original balance is lost. Most of the new added heat remains in the atmosphere for many years into the future.

It is quite likely that it is this additional heat in the atmosphere that is influencing mountain glaciers and polar ice caps.

The second factor in the transformation is the tilted axis of the earth. This causes the intensity of solar radiation to fluctuate from hot to cold; an area is hot where the sun's rays hit the surface of the earth at a nearly ninety-degree angle, while other areas, affected by a considerably lower angle of the sun's rays, experience a net loss of heat and thus get cold. The cool air from the cold areas cools the humid air from the hot areas, and that conflict causes rain.

If oxygen and carbon dioxide are the blood that feeds life on dry land, water vapor is the heart that contracts and expands, causing carbon dioxide and oxygen to flow from place to place. The expansion during water evaporation and its contraction (when cold air meets warm humid air) causes air to move from one place to another and thus nourishes life itself.

As indicated earlier, one liter of liquid water can expand to occupy 1,600 liters of space when transformed to vapor. That is a major displacement of air. Condensation of water vapor is just as dramatic and has the opposite effect to evaporation. For instance, one cubic kilometer of a high concentration of vapor can shrink to one cubic meter of liquid water. That is a major collapse in an open-air environment that would cause air to rush in and fill the vacuum created by such a collapse.

From the reasoning above, the implication of increasing water vapor in the atmosphere is that weather will get more turbulent as the increased vapor production persists. If observations on recent weather storms, typhoons, hurricanes, and tornadoes across the globe are taken into consideration, the conclusion is fairly clear: global warming is mostly a water-vapor problem.

OTHER GREENHOUSE GASES

Methane gas has been listed as one of the other greenhouse gases. Methane and other hydrocarbon gases are created by the breakdown of organic matter, either during decomposition of organic matter or through gastric digestion by animals, such as cattle. Many scientists have been critical of the modern, industrialized production of livestock, claiming centralized large-scale production as a contributive element in global warming.

Other sources of methane and other hydrocarbon gases include industrial mining and forestry carbon decomposition. As in the case of carbon dioxide, methane tends to be heavier than other gases in the air; thus, its insulating contribution is questionable. The exact mechanism by which methane traps heat is not clear. How exactly does a methane molecule deflect or capture heat, and under what circumstances does the molecule release the heat to cause global warming? This is a vital question that begs to be answered by proponents who claim this gas is a factor in global warming.

Nitrous oxide has been listed as one of the other greenhouse gases. Nitrous oxide is formed when lightning streaks through the air, causing nitrogen to combine with oxygen. Increasing levels of the gas in the atmosphere have been observed. However, the exact mechanism by which it traps heat has not been explained well.

POTENTIAL AND LATENT ENERGY

Imagine a rock sitting precariously on the side of a hill. As long as the earth around the rock holds firm, the rock will not move. If, however, the soil around the rock is disturbed, the rock will go into motion, rolling downhill and taking down any obstacles. It will not stop moving until it reaches the bottom of the hill or collides with an immovable obstacle. When the rock is up the hill, scientists describe it as having *potential energy*. The energy can actually be harnessed to perform a task if such a task exists.

Water vapor has two forms of energy: potential energy and latent energy. When water vapor is blown to the top of a mountain, it turns into liquid water (rain) or to solid (snow). That rain ends up in rivers, which, when they spill over a cliff, can turn turbines that generate electricity. Thus, we use potential energy for hydroelectric power.

The *latent energy* in water vapor is heat. This heat must be expended in order for vapor to turn into liquid water. Most of this latent heat does not dissipate into the atmosphere the way long-wave energy does. It is instead passed from molecule to molecule for a long period of time before some of it dissipates.

In science, the law of gravity states that water does not flow uphill. We have become so familiar with the law that we now accept it as fact. In reality, the law is only partially correct. It holds true for liquid water but does not hold for water in the form of vapor. You can prove this to yourself by sitting by a permanent stream. New water will pass by you for hours. You may ask yourself the question, if water does not flow uphill, why doesn't the supply from uphill run out? Common sense should suggest that the water running downhill should eventually run out. The truth is that water in the form of vapor does "flow" uphill. It is the only logical explanation for the perpetual flow downhill. This truism is profound because it means that as much water flows uphill as it does downhill. Water is, in fact, on a treadmill, constantly flowing downhill in liquid form and flowing uphill in the form of vapor.

The point of this discussion is that the atmosphere holds vast quantities of water vapor. This vast quantity exerts considerable influence on air motion, and the shifts in form from vapor to liquid or solid (snow or ice) have profound effects on global temperature. Water can only change form from liquid to gas after absorbing heat energy. As long as it remains in the gas form, water holds that absorbed energy. It must give up that energy to a solid object or other cooler water molecules in order to be transformed back into liquid or solid, as is the case when snow is created. This assertion, though speculative, has serious implications.

THE THREE AMIGOS

Life on our planet is founded on three gasses: hydrogen, oxygen, and carbon dioxide. The unique balance in the interaction among these gases makes life possible here on earth.

Hydrogen

Hydrogen is lighter than air and rarely occurs in pure form. It is highly unstable in that it has a high affinity for oxygen. When the two gases mix, the reaction produces heat, and real fireworks happen. The end result of the fireworks is water—H_2O. So the water is born of fire, but it is also the number-one enemy of fire.

When water is formed during the reaction between oxygen and hydrogen, it shows up as another gas: water vapor. The lightness of the hydrogen atoms greatly influences the behavior of water. When it is heated, the molecule floats like a balloon. This characteristic is a lucky break for us because the floating characteristic of water makes transportation possible for long distances through the air. Without this characteristic, land would be a barren, desolate place.

Oxygen

Oxygen is heavier than hydrogen but lighter than carbon dioxide. It is a very reactive gas. It tends to easily combine with metals, other gases, and liquids to form a wide variety of compounds. As indicated, it combines violently with hydrogen, releasing heat and water vapor.

Virtually all living things, except plants, need oxygen to survive. Oxygen is a major catalyst in allowing those living to burn sugars in our bodies to generate energy for our physical activities. Most of the animal world depends on oxygen to survive. It is likewise a major catalyst in burning fossil fuels to generate the energy we use in automobiles and in industry.

Carbon Dioxide

Carbon dioxide is the heaviest of the three principal gases. It tends to lurk around closer to the ground than the other gases in the air. Oceans are the single largest CO_2 reduction system in the world.

This gas is the nemesis of oxygen. If oxygen promotes a fire to rage, carbon dioxide kills it. In high concentrations, it kills animals.

As oxygen is essential to animals, so carbon dioxide is essential to plants. In the unique partnership between plants and animals, plants absorb carbon dioxide and generate oxygen in their metabolism, while animals absorb oxygen and expel carbon dioxide as waste. It is one of the true wonders of nature, balancing opposed forces to produce harmony.

OTHER GASES

The other principal gases in the atmosphere are nitrogen and water vapor, sulfur oxide, nitrous oxide, methane, and other trace gases.

Nitrogen

According to Wikipedia, the composition of the atmosphere by volume is 78% nitrogen, 21% oxygen, 0.93 argon, 0.04% carbon dioxide, and the remainder is other minor gases. Vapor content varies from 0.4% to 1% nitrogen constitutes 78 percent of the air in the atmosphere; oxygen constitutes 21 percent, carbon dioxide is 0.04 percent and the other gases are 0.98 percent.

Water Vapor

Water Molecule!

You ... mystery of mysteries.
Transparent you are, liquid.
Opaque in dark clouds.
Snow-white in crystal form.
Translucent and slick you are, solid.
You are a gem beyond value.
You shape-shifter, you.
Born of fire yet no friend to it.
Through your changing shapes, you create life
Taken for granted, but you are life itself.

Water vapor is the most dynamic of the atmospheric gases. However, it is pervasive. It can vary in contribution to the volume of air from 0.4% to as much as 1%. Its density changes with temperature, allowing it to soar high when warm and to sink low when cold. Clouds, hovering sometimes as high as 1,000 meters or as high as 10,000 meters (5,000 or 20,000 feet) above ground, attest to the variability and adaptability of water vapor.

Water vapor is by far the most transformative gas in the atmosphere. It holds two forms of energy: potential energy and latent energy. Its potential energy can be witnessed in waterfalls. Most of these waterfalls would not exist if water vapor didn't allow for water to be transported to mountaintops for precipitation and then flow downhill, creating spectacles of beauty. When we gaze at waterfalls, it is impossible to link the water vapor that might have traveled from thousands of kilometers to arrive at locations uphill from the location of the waterfall.

The latent energy is a big mystery. When we gaze at clouds, we know they are soaring high in the sky. We know it is cold up there, due to our common sense and scientific observations. Most of the clouds amass at elevations ranging from 300 meters to 10,000 meters (1,000 to 30,000 feet) above sea level. However, cold as they may be, the clouds are still a huge store of heat energy.

Two considerations might give us a sense of the latent heat energy in the clouds. One consideration is that liquid water absorbs considerable heat energy in order to turn into water vapor. This heat energy does not dissipate until the vapor turns back to liquid water during rainfall.

The second consideration is a comparison; cloud cover floating around at 7,000 meters (20,000 feet) above ground, at a chilly temperature of 5 degrees Celsius, is hot in relation to the −30 degree temperature that might exist at the North Pole at the same time. In short, cold water vapor molecules at 5 degrees Celsius are still capable of transferring considerable heat to even colder molecules at the North Pole.

The transformation of liquid water to vapor is an interesting phenomenon that needs scientific study. It seems that the water molecules get transformed in size instantaneously. If you can imagine the liquid water molecule being the size of a marble, it seems that the molecule transforms to the size of a basketball in an instant.

When water molecules are heated, they vibrate at higher and higher speeds as temperature increases. Molecules in a container vibrate at different speeds and can undergo transformation to vapor individually at various stages of the heating. However, any one molecule may reach a level of vibration where the relationship between the two hydrogen atoms and the single oxygen atom changes dramatically, and at that instant, the molecule size expands. It is probably some kind of quantum leap whereby there is no gradual change in size but a sudden leap from small size to the bigger size after absorbing a critical amount of heat energy. This is a speculative idea, and it is something that would be best pursued by a physicist.

If you watch the bottom of a pot of boiling water, you may witness bubbles develop there, then explode to the surface of the water. The explosive nature of the transformation is best displayed if you try to cook an egg in a microwave. Too many of the water molecules in an egg can reach the critical transformation state at the same time and create an explosion.

The water molecules hold the heat of transformation as long as they are in the form of vapor. The heat is only given up if a vapor molecule comes into contact with a suitable obstacle it can transfer the heat to under the right circumstances.

The Pimp and the Two Suitors

Imagine a pimp standing at a street corner in the red-light district of a town. The pimp is looking for business, mostly hoping to take advantage of lonely people and hook them up with his overworked and underpaid sex slaves. The pimp runs into his first prospect, who we can call C.O.Two in order to protect his identity. He says to C.O.Two, "You look lonely and sad. I can hook you up with hot companions who will give you excitement you have never experienced before." C.O.Two is reluctant, but agrees to a blind date with potentially exciting escorts at a rendezvous location. He moves on, still excited about the prospects of meeting a hot escort.

The pimp stays at his corner waiting for his next victim. Again, for reasons of confidentiality, we will call her H.Two O. The pimp says to H.Two O, "The two of us go back a long time. We have been close since we were kids. I think I have found a perfect match for you. This guy is wild, but he is also fairly cool. He seems like the sort of guy who will make you happy forever." H.Two O does not trust pimps generally but knows this particular pimp quite well and thinks he might have an interesting prospect. She agrees to the plan for the three of them to meet at the special rendezvous location.

At the scheduled meeting place, H.Two O and the pimp arrive at the same time and are a bit earlier than the scheduled time. In light conversation, H.Two O asks the pimp about the prospect. The pimp, trying to avoid answering the question directly, says, "C.O.Two might drop by." H.Two O's heart sinks. She has met C.O.Two many times before; there were no major sparks between them. But lacking any other plan, she decides to hang around. C.O.Two arrives a bit late but is dressed in his best.

There is a brief exchange of pleasantries but still no sparks. The pimp then pulls a magic wand from his pocket and waves it in a striking motion. Instantly, there is a bolt of lightning, and before the onlookers realize what has happened, white stuff has spilled all over the place.

Now, it is not what you think. The white stuff is starch; the pimp is chlorophyll. The lightning is pockets of light from the sun called photons. The two blind dates are carbon dioxide and water. The point of this story is that carbon dioxide and water are always hanging around together to form carbon-containing compounds. Once the carbon-containing compounds break up, both are released at the same time. If you observe the release of one of them increasing, there is an equivalent release of the other at the same time.

The reaction between carbon dioxide (C.O.Two) and water (H.Two O) is the fundamental reaction that creates all the energy we generate from fossil fuels and the sugars we burn in our bodies to generate energy. Invariably, the by-products of our bodies burning sugar as well as industries burning fossil fuels are carbon dioxide and water in the form of vapor.

From the initial starch, plants fabricate sugars, proteins, and solid- wood fiber. Animals consume plant materials and generate fats and oils that eventually become petroleum, natural gases, and other compounds. In this sense, all fossil fuels originate from one common source, a photosynthetic reaction in which carbon dioxide and water combine with the help of sunlight.

Highlighting carbon dioxide as the key greenhouse gas in the greenhouse gas theory is an idea that has some weaknesses.

THE WEAKNESSES OF THE GREENHOUSE GAS THEORY

Greenhouses are used to grow crops such as tomatoes and lettuce under controlled conditions. Greenhouses consist of a framed building with transparent covering. Sunlight penetrates through the transparent covering to get to the plants inside. The plants and the soil absorb the sunlight and radiate some of the absorbed energy as heat.

Quite often, greenhouses ventilate either through controlled openings or by fanning air around. The ventilation is necessary to avoid the buildup of water vapor and to ensure that too much heat doesn't too negatively affect the plants. The key elements of the greenhouse are the outer shell that traps the air inside, the trapped air, and controlled ventilation.

A blanket on a bed works rather like the outer shell of the greenhouse. Again, the purpose of the blanket is to trap air around a person sleeping underneath. The body of the covered person generates the heat, and the trapped air warms up to keep the person comfortable.

Again, the key characteristics of a blanket are the ability to trap air and keep it reasonably stationary. Other than the occasional breathing in and out that causes a few waves, the movement of air in and out of the blanket is controlled. The air still moves around in a controlled fashion.

The greenhouse and blanket are examples provided to show how small quantities of air are heated by an energy source. In the case of the greenhouse, most of the heat is provided by the sun. For the blanket, the heat is provided by the covered body.

At a global level on our planet, the situation is slightly different. By and large, the air is not confined. It moves depending on atmospheric conditions. The source of heat for our planet is the sun. The sun's radiation comes in as waves of shortwave bursts of energy. The soil, rocks, and other objects absorb the sun's energy and convert it into long-wave bursts of energy—heat.

The proponents of the greenhouse gas theory maintain that carbon dioxide and the other greenhouse gases capture both the incoming and the outgoing energy, hold it for a while, and release it later, creating a progressively warming effect on the atmosphere.

There are several weaknesses in the arguments supporting the carbon dioxide based greenhouse gas theory. These are explained in the following sections.

Carbon dioxide is heavier than air. Carbon dioxide, deemed as the key offender in the greenhouse theory, is heavier than air. This means that its highest concentrations are close to the ground in low-lying areas. Without the benefit of experiments to the prove the point, it can still be postulated that the highest concentrations of carbon dioxide are within 1,000 meters (3,500 feet) above sea level, and concentrations will diminish as air climbs higher into the atmosphere. Incidentally, this is not unusual. Oxygen is known to diminish with elevation as well.

Since carbon dioxide is most prevalent when close to ground level, it is unlikely that it would absorb significant levels of ultraviolet light and generate outgoing, long wavelength radiation (heat) and hold enough of it to influence global temperatures.

Carbon dioxide is a very stable gas. Carbon dioxide will change form only if it is subjected to extreme conditions. It will turn to liquid if it is compressed to 350 psi (pounds per square inch). It will become solid if it is cooled to −80 degrees Celsius. Both conditions rarely exist in normal atmospheric conditions.

Given the stable nature of carbon dioxide, it is difficult to rationalize the claim that it regulates vapor through absorbing ultraviolet light. Technically, vapor floats above carbon dioxide because they have different densities. So how exactly does carbon dioxide regulate vapor? This is not well explained in literature.

The proportion of carbon dioxide in the atmosphere is the lowest of the three major gases in the atmosphere. The key gases are nitrogen (78 percent), oxygen (21 percent), carbon dioxide (0.04 percent) and other gases such as methane, argon, nitrous oxide, which constitute the remaining less than 1 percent.

Although the amount of carbon dioxide in the atmosphere is increasing substantially, it is unlikely that environmental disasters are caused by its increase.

Both carbon dioxide and water vapor are usually released simultaneously at high temperatures during combustion or metabolism, as when air comes out of an exhaust pipe. It is therefore difficult to justify the claim that carbon dioxide regulates levels of water vapor in the atmosphere. Neither of them really regulates the other. They are independent. In fact, there are more situations where water vapor is released without carbon dioxide being involved in any way. For instance, when the sun heats rain-soaked soil, water vapor is created; however, carbon dioxide has absolutely no part to play in that evaporation.

The theory does not fully explain the magnitude of the damages. It is becoming apparent that levels of precipitation in the form of snow and rain are increasing erratically. News of catastrophic floods has become very common. During the winter of 2007, historic snowfalls in China shut down transportation systems and caused many deaths.

In the United States in 2008, besides the major hurricanes, the upper Mississippi basin was inundated with historic rainfalls that lasted more than a week in some local areas. The rains caused major flooding and displacement of people and livestock.

In Canada, severe weather events have increased over the past ten years. These include Hurricane Igor on September 21,

2010; the Manitoba and Saskatchewan floods in September 2011; the Richelieu, Quebec, floods of spring 2011; the Goderich, Ontario, tornado in August 2011; the British Columbia floods of spring 2012; the Alberta flood emergency in 2013; and the eastern Canada ice storm of 2013.

These major precipitation catastrophes cannot be explained by the carbon dioxide-driven global warming theory. They are likely to become more common.

In the Canadian coastal mountains and in the Rockies, ice caps and glaciers are retreating at a phenomenal rate. The snowcaps on Mount Kilimanjaro, Mount Kenya, and on many mountains around the world are disappearing rapidly.

Besides retreating glaciers, the North and South Pole ice caps are also being affected dramatically. The magnitude of change is not commensurate with the increasing quantities of carbon dioxide.

Misleading Statistics

Statistics can be a very useful tool in explaining behavior and rationalizing scientific phenomena. However, statistical information can also be misinterpreted or even misused. The IPCC report titled "Climate Change 2007—Synthesis Report" states that the initial theory under consideration was that water vapor was probably the primary greenhouse gas. After some consideration, this theory was discounted. The reason for dropping water vapor was that the increase in carbon dioxide over the past century seemed to be more significant than the increase in water vapor. Analysis of the statistics indicated that carbon dioxide had increased by 35 percent compared to an increase of only 8 percent for water vapor.

On the surface, the IPCC conclusion seems reasonable. However comparison of percentages can be misleading. An example using more familiar transactions might help to explain the dilemma. Consider two people, one Mr. Poor and the other

Mr. Rich. Mr. Poor has a savings account of $1,000. Mr. Rich has a savings account of $10,000. The two men have a mutual friend, Mr. Donor, who deposits $100 into the savings accounts of each of them because they are equally valuable to him.

From the perspective of the accounts of the two men, Mr. Poor's account increased by 10 percent and Mr. Rich's increased by only 1 percent. This seems fair because Mr. Rich is already loaded.

If we apply the IPCC percentages as annual interest rates on the two savings accounts, we will find that 35 percent applied to $1,000 will yield $350 in the first year for Mr. Poor while 8 percent will yield $800 for Mr. Rich. So, in absolute terms, Mr. Rich is gaining more with an 8 percent interest rate than Mr. Poor with an interest rate of 35 percent.

The point of this example is that carbon dioxide constitutes less than 0.05 percent of the gases in the atmosphere, while water vapor constitutes 0.4 and 1.0 percent of the gases in the atmosphere by volume. So a modest 8 percent increase in its quantity may be quite influential in affecting global warming.

Another dodgy statistical interpretation is in drawing inferences from relationships between two attributes. For instance, the IPCC has relied quite strongly on the fact that a plot of carbon dioxide levels against global average temperature increase shows an exponential trend.

However, we must consider that during the combustion of fossil fuel, both carbon dioxide and water vapor are released simultaneously. Analysis, then, of the relationship between carbon dioxide levels and global warming should attempt to remove the effect of water vapor in the relationship. If the water vapor is included as a covariate, the effect of carbon dioxide on global warming might look quite different.

Secondly, a relationship may exist between two factors, with one being the cause of or a reaction to the other. For example, in biology it is known that for most human beings, there is a linear relationship between arm length and height of a person. A five-foot person has arms that are proportionally shorter than the arms of a six-foot person. But we cannot draw cause-effect conclusions from this relationship. Both are a result of normal proportional growth.

Below is an example of US fuel consumption. In the figure, it is clear that US fuel consumption has increased exponentially since 1775. If we were to correlate the US petroleum consumption against global increase in temperature, we might conclude wrongly that the US petroleum consumption alone has caused global warming, because the rise in fuel consumption there seems to parallel the trend in increasing global temperatures.

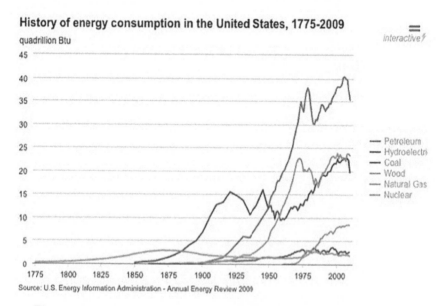

Figure 20. History of fuel consumption in the United States

AN ALTERNATIVE THEORY ON GLOBAL WARMING

The genesis of all fossil fuels was described briefly in the story of the pimp (chlorophyll, C.O.Two, H.Two O, and energy from the sun). It is the story of how chlorophyll takes water (a liquid), carbon dioxide (a gas), and photons, or bursts of energy from the sun, to create solid matter—starch.

Starch is the foundation of all fossil fuels, plant and animal foods, and beverages, such as alcohol and sodas. In this profound process, all life depends on this compound; without it, life would cease to exist. It is also important to note that water is essential to forming carbohydrates and thus for supporting life. As an illustration of the importance of water, if you go to the middle of the Sahara Desert, you will find lots of carbon dioxide, lots of oxygen, and there is more than enough sunlight there. However, the number of living things there is limited. The lacking ingredient is water.

If we accept the profound reality of photosynthesis, it becomes clear that at the beginning of all fossil fuels, there is carbon dioxide and water. It should, therefore, be expected that during the combustion of fossil fuel, both carbon dioxide and water are released. Essentially, combustion reduces fuel to the original basic elements. This will be explored in more detail later. In addition to carbon dioxide and water, combustion also releases heat.

In its most basic form, the reaction between water and carbon dioxide takes the following form:

$$2CO_2 + 2H_2O \rightarrow \text{Carbohydrate}$$

In essence, the sun's energy helps to bind carbon, oxygen and hydrogen together in a special way to produce starch. For Plants to produce proteins, fats, sugars, and carbohydrates further modify this initial compound produced by photosynthesis.

Animals feed on plants and further transform the foods originating from starch into even more by-products.

When plants die and are buried underground, they eventually become coal. When animals on land or in seas die and are eventually buried, their fats become petroleum. From this, we derive a wide variety of fuels.

The essential message is that fossil fuels are actually water and carbon dioxide that have been sequestered for hundreds or perhaps millions of years. When we burn these fuels, we release the initial compounds that were used to create them.

The observation of increasing levels of carbon dioxide indicates that significant combustion has occurred, thus causing the release of heat, water vapor, and carbon dioxide. It may be misleading to focus solely on carbon dioxide and then conclude that it is the key factor in global warming. The nature of the combustion process itself and the unstable nature of water in regard to varying temperatures may, in fact, be the pivotal factors in global warming.

Water tends to transform from solid to liquid to gas (vapor) within normal atmospheric temperatures (−30°C to +40°C). Heat energy is required for the transformation of solid water to liquid and of liquid water to vapor. In essence, water vapor is a reservoir of considerable heat energy.

Another significant characteristic of water is that it changes density due to changes in temperature. The transformation from water to vapor results in the expansion of space occupied by water so that its density reduces dramatically to the extent that hot water vapor is lighter than air.

Surprisingly, the transformation from liquid to solid also results in the expansion of space occupied by water molecules. This makes ice lighter than liquid water.

The shape-shifting and density-changing characteristics of water make the transportation of water possible to areas that would otherwise have been barren. As a vapor, water flows uphill.

The transformation from liquid to ice has so much force that it can shatter large rocks, reducing them to rubble.

THE IMPACT OF MODERN SOCIETY ON WATER VAPOR LEVELS

Combustion

The production of water vapor from combustion has already been described. Combustion occurs in various forms. The most obvious examples include internal combustion engines and the burning of fossil fuels to generate heat for domestic and industrial use.

Figure 21. Means of transportation
(Photo by Dr. Sam Otukol)

One overlooked form of combustion is the metabolism of carbohydrates and sugars by animals and plants to supply energy for their survival. In animals, the metabolism process generates both carbon dioxide and water vapor. Most of these gases are discharged through exhaling. But the gases can also be expelled through the skin. Over the past fifty years, the world's population has more than doubled. By now, the amount of water vapor generated by our activities has increased substantially. One billion people release less water vapor than six billion people. When one considers the increase in animal populations over the past fifty years as well, it can be concluded that animal metabolism is as much a problem as is the increasing use of automobiles.

In plants, the effect of manufacturing organic matter (photosynthesis) is a production of oxygen. While oxygen is discharged through pores in the leaves, so is water vapor.

Cooking and Boiling

The acts of cooking food and producing boiled water generate considerable amounts of water vapor. Virtually no modern family neglects to boil water to make coffee and tea or to cook food. Compared to a hundred years ago, there are now many more events for which water is boiled. Also, there is a significant increase in the use of gases such as butane and methane for cooking and domestic heating. These gases generate more water vapor than carbon dioxide by volume, though both are generated at the same time.

Moreover, more families now use warm water to bathe than they did one hundred years ago. A 10-minute shower taken by 50 people, can release more water vapor into the atmosphere than a modern efficient bus traveling 4 kilometers in a downtown area. One person using 50 liters of warm water to shower releases close to one liter of water as vapor. An efficient bus burns 1 liter of fuel every for kilometers, and that fuels produces close to the equivalent of one liter of water as vapor. So 50 people showering in warm water can do more damage to the environment than a bus travelling 10 kilometers downtown.

Generation of Nuclear Energy

Most discussions of global warming tend to paint nuclear energy as a clean energy source, created by humans without harm to humans. In reality, it takes huge volumes of water to cool nuclear reactors. The superheated water is normally discharged into lakes or into the oceans.

Figure 22. Nuclear energy generation

The harmful nature of discharging hot water into lakes is not easily apparent. But in recent history we have heard much about El Niño, the unusually warm water currents that sometimes influence weather dramatically. El Niños were rare thirty to fifty years ago, but today we experience them at least every two to four years.

The discharge of hot water into lakes and oceans upsets the temperature equilibrium of the affected waters. This, in turn, increases evaporation, thus affecting the balances in the atmospheric quantity of water and the amount of energy stored by vapor.

The invention of the internal combustion engine (1806) may have set us on a path of increasing the generation of CO_2 and H_2O. Every person on earth today aspires to own an automobile for freedom of movement and convenience.

The world's population is growing at the rate of 3 percent per year, and several countries' economies are just coming to life. It is likely that the use of family-owned automobiles will triple in the next thirty years. The implications of this development are bound to be significant with regard to global warming.

Irrigation

As the world population has increased, modern agricultural practices have come to rely increasingly on irrigation to ensure consistent agricultural output. It is not the first time on earth that population density has caused an increase in the use of irrigation to produce crops. The Egyptian civilization as well as the Mayan civilization relied on irrigation. To our amazement, these two civilizations seem to have collapsed in spite of their considerable advancement. Could it be that increased water vapor from evaporation affected their rainfall patterns? There are no living or written witnesses to testify to this, but it is something to consider.

The conversion of liquid water to vapor during the crop-growing season could lead to a permanent cycle of convection currents, sending hot water vapor upward and drawing in dry cool air. If the cycle coincides with the wet season, then regular rainfall could diminish and cause prolonged droughts.

The extensive nature of the Sahara Desert suggests that something dramatic must have happened to cause such a dramatic change in the local climate. From biblical stories, it is clear that Egypt was the most bountiful producer of agricultural crops at some point in history.

Figure 23. Crop irrigation

In one story, Joseph, the son of Jacob, was sold by his brothers to Egyptian merchants. Over time, Joseph became influential in the Egyptian pharaoh's house and was made custodian of food stores.

Meanwhile, in Israel, a bad drought started, and the people ran out of food. Joseph's brothers, who had sold him into slavery, eventually came calling for assistance. Joseph did eventually rescue his family, but in the story, Joseph indicates that the drought was expected to last five more years after the time his brothers arrived. This suggests that the Egyptians had knowledge of the cycles of drought.

If the Egyptians were experiencing the same drought that the Hebrews were experiencing, how was it that they still had considerable reserves of food? The answer may be that they were farming the area west of the Nile extensively and were using irrigation to produce their crops. To this day, Egypt depends almost exclusively on irrigation to produce food. Could it be that these descriptions in the Bible included the makings of global warming at that time?

How Heat Is Balanced In The Atmosphere

The atmosphere consists of 0.4 to 1.0 percent water vapor by volume. At any given point, the level of water vapor may rise or fall depending on sources of water evaporation or ambient temperature.

Cool temperatures tend to reduce the volume of air and correspondingly increase the relative volume of water. When temperature increases, air volume increases and accordingly reduces pressure where the heat is located.

The holistic way to try to mine the reasons for global warming requires a close examination of the way that fossil fuels are created and then observing what happens when they combust.

The view in this book is that all fossil fuels on earth start with photosynthesis. The first by-product of this process is starch. During the creation of starch, four elements must be present: carbon dioxide, water, chlorophyll, and a burst of sun energy. The sun energy powers the fusion of carbon, hydrogen, and oxygen into new molecules that have the capacity to supply energy for life or to power machines as may be required.

The initial product (starch) is modified to create sugars, carbohydrates, proteins, and fats. From the sugars, alcohol can be synthesized. From the fats, various forms of oils can be formed.

There is a similarity between the structures of the molecules that power biological processes and the structure of fossil fuels. Some examples are provided below.

Table 1. Products that power animal and plant life

Sugar and Carbohydrates	Chemical Form	By-Products When Carbohydrates Are Burned during Respiration
Sucrose	$C_{12}H_{22}O_{11}$	12 molecules of carbon dioxide plus 11 molecules of water
Glucose	$C_6H_{12}O_6$	6 molecules of carbon dioxide plus 6 molecules of water
Fructose	$C_6H_{12}O_6$	Same as above

Table 2. Alcohols

Product	Chemical Form	By-Product after Burning
Ethanol	C2H5OH C2H5OH+3O2→2CO2+3H2O	2 carbon dioxide molecules plus 3 water molecules
Methanol	CH3OH 2CH3OH+3O2→2CO2+4H2O	1 carbon dioxide molecule plus 2 water molecules

Table 3. Hydrocarbon gases

Gases Hydrocarbon	Composition	By-Products after Burning
Methane	CH4	1 carbon dioxide molecule plus 2 water molecules
Propane	C3H8	3 molecules of carbon dioxide plus 4 molecules of water
Butane	C4H10	4 carbon dioxide molecules plus 5 water molecules
Acetylene	C2H2	2 carbon dioxide molecules plus 1 water molecule
Ethylene	C2H4	2 molecules of carbon dioxide plus 2 molecules of water

Table 4. Fossil fuels

Automobile and Aircraft Fuels	Composition	By-Product after Burning
Diesel	75% saturated hydrocarbons and 25% aromatic hydrocarbons	Some carbon produced as soot, carbon monoxide, and water vapor

Petrol	Consists of a mix of hydrocarbons, including:	
	a) Octane C8H18	8 carbon dioxide molecules plus 9 water molecules
	b) Pentane C5H12	5 carbon dioxide molecules plus 6 water molecules
	c) Nonane C9H20	9 carbon dioxide molecules plus 10 water molecules
	d) Hexadecane C16H34	16 carbon dioxide molecules plus 17 water molecules

Table 5. Household fuels

Common Fuel Names	Composition	By-Products
Kerosene	C12H26	12 molecules of carbon dioxide plus 13 molecules of water
Paraffin wax	C20H42	20 carbon dioxide molecules plus 21 water molecules
Charcoal	(Cellulose) C6H10O5	6 molecules of carbon dioxide plus 5 molecules of water
Coal	Same as above	Same as above

An examination of the by-products of burning most fossil fuels reveals that the key by-products of metabolizing or burning hydrocarbon-based energy sources are carbon dioxide and water. Since heat is involved in these processes, the water produced as a by- product is released as water vapor, and as such it contains the heat of transformation by default. For animals, the vapor from metabolism is released from the lungs. For plants, the vapor is released as evapotranspiration. In fact, it seems that for every single molecule of energy source burned, more water molecules are generated than carbon dioxide molecules.

Essentially, when carbon dioxide is produced from fossil fuels, water vapor is also released simultaneously. The two gases are linked inextricably, and a measurement of the increasing level of one of them implies that the level of the other is also increasing.

It is easier to measure levels of carbon dioxide by taking air samples from different fixed locations and at regular intervals. The levels of water vapor, on the other hand, are difficult to measure since the water molecules can easily change form—from vapor gas to liquid or solid. Secondly, the gas tends to change density dramatically when temperatures are affected.

It requires a considerable amount of energy to convert liquid water into a vapor form. At the boiling point, more than 1,200 calories of energy are required to transform one liter of water into vapor. What this means is that the water vapor in the atmosphere is actually a reservoir of heat energy. If the vapor in the atmosphere is converted back to liquid water, the transformation only occurs when the relatively warm vapor transfers its heat to colder vapor molecules.

Most of the heat exchange occurs between water molecules, so when there is an increasing supply of water vapor molecules, the heat exchange results in increasing global temperatures. If the supply of water vapor is constant, equilibrium is reached where temperatures are stable.

The transformation of liquid water to vapor is an area of science that needs to be explored. Generally, we describe the process as evaporation. However, the actual transformation at the molecular level is an instantaneous shift from one form to another. It could probably be described as a quantum leap.

Weather and Climate

Climate is a dance of water vapor
Choreographed by God.
Moisture rising uncontrolled,
Dashing left then right,
Searching for a home east then north,
Guided by an invisible hand.
Coolness rushing to warmth;
Warmth seeking coolness.
Transformation; liquid to gas then back again,
Seeking balance where none can be found.

Weather is really a dance of water vapor, perhaps choreographed by God Himself. The dramatic shift in the density of water vapor with changes in temperature is what causes air in the atmosphere to move about. Where it is hot, water vapor is light and thus rises to higher altitudes. This rise causes an air deficit at the hot spot, and colder, denser water vapor rushes to the hot spot.

The hot and cold movements cause turbulence. The upward movement of hot air is actually quite dramatic and a well-known phenomenon to airplane pilots. The effect of the water transformation is so powerful that commercial aircraft have to fly 7,600 meters to 10,000 meters (25,000 to 35,000 feet) above ground to avoid the turbulence. The turbulence is caused by a transformation of water from liquid to vapor. The transformation is not different from the explosion of dynamite when solid explosive material transforms into a gas.

The sun's energy is needed to transform liquid water to gas (water vapor). When the transformation occurs, most of the vapor rises to an altitude of 3,000 to 15,000 feet. Considerable energy (heat) is required to transform the water and cause the vapor to raise that high.

An average commercial airplane with two hundred passengers requires 10,000 liters of fuel to take off and rise to an altitude of 6,100 meters (20,000 feet). The total weight of the aircraft, fuel, and passengers is 40 to 50 tons. On warm, sunny days, hundreds of thousands of tons of water are transformed from liquid to vapor.

The vapor carries heat energy. When hot humid air rises, it encounters cooler air, and a heat exchange occurs. The cooler air gets warmer; the warmer air gets cooler. This heat exchange ultimately results in the recreation of liquid water and rain droplets (or snow). Liquid water is heavier than air, so rainfall results. On a global scale, if the amount of water evaporated or converted to vapor through combustion annually were stable, temperature equilibrium would be achieved. The global temperature would be generally constant. However, if the amount of water evaporated on an annual basis were increased, the average air temperature would increase as well. The heat that caused water evaporation is simply exchanged, and it does not necessarily dissipate into the atmosphere, as does long-wave radiation from soil.

From the big-picture perspective, the impact of carbon dioxide should not be totally discounted in the climate change process. From a narrow perspective, it is tempting to confine its role to that of an indicator variable, and in that role it is vital for giving us bearings on where we are in terms of worsening global warming. It so happens that most processes that generate water vapor also generate carbon dioxide, and therefore, tracking fluctuations in carbon dioxide is a good way to keep pace with or predict what might happen next as we continue industrialization.

The trend toward increased and catastrophic precipitation is bound to get worse. The use of gas to heat homes, the use of automobiles, and the need to use fossil fuels to facilitate development are all bound to increase.

Tragically, technologies that are likely to increase water evaporation are aggressively promoted as clean-energy sources. For instance, some proponents are advocating the increased use of hydrogen- powered cars. Guess what? The exhaust from a hydrogen-powered car is pure water vapor!

A further example of risky activities that could result in tragedy is the increasing use of irrigation to produce food crops. This is an essential activity for the world's growing population. But irrigation produces the equivalent of a hydrogen bomb effect in terms of transforming liquid water to vapor. In the long term, excessive use of irrigation can cause permanent air-turbulence effects that could lead to land desertification.

Using the theory of water transformation causing adverse weather effects, it might be useful to explore links to desertification. One factor that could aggravate the effects of evaporation from irrigated fields is the proximity to the Tropics of Capricorn and Cancer. In the summer, the Tropics of Capricorn and Cancer receive the most intense concentration of solar radiation, with several days of near vertical sunrays focused on the same area. The Tropics of Cancer and Capricorn are the turning points where the movement of vertical sunrays overhead, toward the south or north, stop and reverse direction.

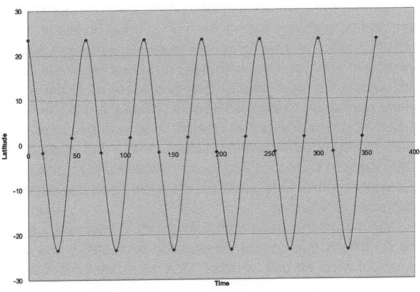

Figure 24. Sun's movements from the
Tropic of Cancer to the Tropic of Capricorn

Africa is the continent with the most land coverage between the Tropics of Cancer and Capricorn; it also happens to have the severest desert conditions. Both the Kalahari (Namibia) Desert and the Sahara Desert fall along a 1000-kilometer band on either side of the Tropics of Cancer and Capricorn. This might be merely a coincidence, but it could also be a correlation between the tropic lines and desertification. A further observation is that the Australian Outback desert, the southwestern part of the United States, and the deserts in India and Mexico fall within the Tropics' 1,000-kilometer band.

How Does Weather Work?

Geographers and climatologists have been studying weather for more than two centuries. Even before ancient societies noted changes in temperature over time, they developed methods to foretell when the rains would come and when to plant crops.

General principles of agriculture indicate that after sunset, land and bare ground cool down quickly while bodies of water (oceans, lakes, and seas) remain relatively warmer. It is also known that warm air rises, and cooler air moves below it to fill the void. During the night, cool air over land will move to the ocean where warmer air has moved upward. When the sun rises over land, the earth warms quickly, creating an upward air motion. This cyclic motion of air from water bodies to land and back allows water vapor to be moved from ocean lakes and seas onto land. The water vapor is only beneficial if it cools down sufficiently to be precipitated as water.

The movements of air from cold places to warm places and the conditions that control atmospheric temperature are what constitute what we call "weather." The air movements result in storms, rain, snow, clouds, and fluctuations in daily temperatures.

Weather, then, is the daily record or fluctuation of temperature, humidity, rainfall, and wind. In some locations on earth, the ideal movements of air can quite often be violent, consisting of tornadoes, gale winds, and thunderstorms. In other places, the air movements are gentle, and there is calmness throughout the year.

The strength of air movement is a testament to the dramatic nature of the conversion of water-to-water vapor. Although no loud explosive noises are heard during the transformation, the effect can be dramatic. For instance, daytime commercial planes must fly at 25,000 to 35,000 feet above sea level to avoid air turbulence that would be experienced at lower elevations. It is reasonable to conclude that prolonged substantial transformation of water from liquid to gas may cause long- term or permanent airflow patterns; these patterns can change if the underlying source of the transformation is removed or altered.

It is reasonable to speculate that prolonged and substantial transformation of liquid water-to-water vapor on an exposed soil surface would lead to the equivalent of air expansion associated with a hydrogen bomb explosion.

In this concept of air moving from a body of water (lake, ocean, or sea), let's assume that some level of balance has been established over a land area. The area is covered in thick forest. At night, the land is relatively cool, and the lake or ocean is warmer. Air moves from the land area toward the lake (or ocean).

In the morning, the sun rises and quickly warms the land, but the lake or ocean is relatively colder. So air moves from the lake toward land.

On land, the thick forest generates evaporation and transportation. The air coming from the ocean is humid, having evaporated all night. The humid air from the ocean mingles with humid air from the forest. At a suitable point, the right elements meet to cause precipitation. These elements are very warm humid air (relatively) colliding with colder air at the cloud level, and a relatively stable temperature at the ground level.

The combination of evapotranspiration from the forest and humidity from the ocean (or lake) ensures that a high humidity occurs. The rising air cools down as it mingles with even colder air at high altitude levels. Precipitation results if sufficient heat is removed from the rising humid air.

The evapotranspiration is controlled because the temperature at ground level is stable. In addition, the trees shut down transpiration to avoid losing too much moisture.

If, however, the forest is cut down and agriculture is established, the exposed soil warms up much more than the forestland. The transformation from water-to-water vapor is explosive because soil temperatures can rise as high as 300 degrees Celsius. The air over the land gets quite warm, resulting in lower relative humidity.

Under the modified ground conditions, the chances are reduced of humidity rising sufficiently to cause easy precipitation. Although more moisture is released into the atmosphere, the location where precipitation occurs is less predictable due to turbulence. The modified ground conditions may now cause unexpected droughts because the air turbulence has changed the conventional airflow patterns.

The explosive nature of the transformation of liquid water-to-water vapor is evident in two forms. First, observation of early morning clouds will show that they are mushroom shaped. This mushroom shape is now familiar to us because we have seen enough open-air nuclear blast photos and videos to recognize the shape.

Second, evidence of the violent nature of water transformation is the air turbulence that can be experienced in an aircraft flying at a relatively low altitude over land. The air turbulence can be violent enough to bring down an aircraft.

It is fair to conclude that conditions can be created at the ground level where the effects of the water transformation can be equivalent to a hydrogen bomb blast, and if the conditions are maintained for long enough, the transformation can lead to a permanent alteration of weather conditions in a local area.

How Does Climate Happen?

If we interpret weather as being the sum of local weather events over the long term, then we could interpret climate as consisting of events that occur from season to season, over a period of a year or longer.

Climate is actually influenced mostly by the tilt of the earth's imaginary axis that runs through the North and South Poles. The tilt of the axis is 23.5° from the vertical position or orientation. Further references to this will round the tilt down to twenty-three degrees.

To envision this, imagine that the earth is shaped like an American football. One pointy end is at the North Pole, and the other pointy end is at the South Pole. The pointy ends are distorted. Instead of the imaginary axis running through them, forming a ninety-degree angle with the axis that runs through the center of the sun and the center of the earth, the axis is tilted about twenty-three degrees toward or away from the sun.

Although very few people really know about this angle, it affects our climate more than anything else. Over a one-year period, this angle makes it seem that the sun is traveling north and south on a regular cycle. For example, we expect the sun to be north in June and south in December. This is all attributed to what can be described as the "angle of grace." This tilt defines when people north of the equator will have summer, autumn, spring, and winter.

To explain the significance of this angle of tilt, imagine the earth's axis forming a ninety-degree angle with the axis running through the center of the sun and the center of the earth. If this were the case, there would be no seasons. The hottest area on earth would be at the equator. Temperatures would cool off as you moved toward the poles. Every single day's daylight would be the same length as the day before; this is how it would be all year long. The earth would look very different from the way it is now. This will be explained in the next section.

The 1,000-Kilometer Beam from the Sun

Imagine the sun as a flashlight, able to illuminate a circle 1,000 kilometers wide. Let's say the most important property of the beam of light is that the rays of light hit the earth's surface at approximately ninety degrees. Due to the earth's tilted angle, the beam of light would move up and down between the Tropic of Cancer and the Tropic of Capricorn over a period of one year.

Around June 22, the beam will be at the Tropic of Cancer (north of the equator). Around September 22 and March 22, it will be at the equator. Around December 22, it will be at the Tropic of Capricorn (south of the equator). Around March 15, it will be back at the equator, and the cycle would continue.

The 1,000-kilometer beam of light from the sun travels at a steady pace, moving forward 150 kilometers every day. Every square foot of land in the area between the Tropics of Cancer and Capricorn is visited by this beam of light for one day during the year.

One important characteristic of this beam of light is that it lingers a bit longer at the two tropic parallels. At these latitude, the beam would move forward one day, stop the next day, and start moving in the opposite direction the next day.

Although the time taken by the beam of light from the sun during the reversal of direction seems to be short, it may be that this short time has influence on the climate along the Tropics of Cancer and Capricorn.

The 1,000-kilometer beam of light from the sun has been used here to illustrate a point: the rays of light from the sun cover half of the earth's surface at any given time. However, in areas where the sun's rays form a ninety-degree angle with the earth's surface, there is a higher concentration of energy per square foot than when the angle of the sun's rays is smaller than ninety degrees.

The beam-of-light concept happens in reality. Its radius is obviously bigger than the 1,000 kilometers used in this example, but the ninety-degree angle of the sun's rays is realistic. Of course, the local topographies affect the influence of this angle, but ultimately, the concentration of sun energy in the beam is a real effect.

The most significant evidence alluding to the effects of the beam of light is that most of the deserts of the world lie within the proximity of the two parallels (the Tropics of Cancer and

Capricorn). The phenomenon is most obvious in Africa. The Sahara Desert lies squarely along the Tropic of Cancer. In South Africa, we have both the Kalahari and the Namibia Deserts. Both are located within the bands of land along the Tropic of Capricorn.

In North America, the Mexican deserts and the semiarid lands in California all fall within the Tropic of Cancer band.

In Australia, the Great Sandy Desert, Gibson Desert, and Great Victoria Desert all lie along the Tropic of Capricorn belt. Farther inland, you find the Simpson Desert falls along the Tropic of Capricorn corridor.

Most of the deserts in Middle Eastern countries (Israel, Saudi Arabia, Egypt, Libya, Kuwait, Jordan, etc.) all lie within the Tropic of Cancer belt.

Pakistan, Afghanistan, and western India are affected, to some extent, by the effect of the Tropic of Cancer. Asia is saved from the more adverse consequences of the Tropic of Cancer effect by the Himalaya Mountains and associated mountains of central Asia.

The significance of these two Tropics is that they are net sources of heat for promoting water evaporation from oceans and lakes. They do this by raising general air temperatures, thus increasing the capacity of air to hold more water vapor. When liquid water absorbs the heat and turns to vapor, the heat gets locked into the water molecules as transformation energy. In this form, the heat energy can circle the globe for several weeks before it is finally dissipated, when the vapor is transformed back into liquid water.

In South America, the area traversed by the Tropic of Capricorn is relatively less affected by the desertification that has plagued the other continents. The only evidence of the effect is the relatively arid land on the west coast of Chile.

The orientation of the ring of mountains along the west of the South American subcontinent plays a significant role in preventing more extensive desertification. The effect of the mountains occurs in two forms.

First, the elevation of the land moderates the angle at which the sun strikes the surface of the land. The elevation of the land has the net effect of spreading out the surface of the land; thus, the energy absorbed per square foot of surface area is less than what would occur on flat land.

Second, the elevation of the land also cools the air, preventing excessive heating. As indicated earlier, cool air has less capacity to hold water vapor than hot air. So a cool breeze blowing over a lake will cause less water evaporation than a hot breeze blowing over the same lake.

The only other large landmass in the Southern Hemisphere that is traversed by the Tropic of Capricorn is Australia. As in Africa, Australia has major desertification problems. Almost the entire western half of Australia is desert, and the desertification is spreading.

The Chicken-on-a-Bad-Rotisserie Effect

Besides the earth rotating around its tilted axis, it is also revolving around the sun. Here, again, we see what seems like a mistake in nature. The orbit of the earth around the sun is not a regular circle. It is an ellipse. The orbit is elongated like a stretched-out elastic band. This means that at some times of the year, the earth is closer to the sun than at other times. Physicists can explain why this happens; all we need to know is that it happens.

The earth is closest to the sun around June and December and farthest around March and September. The advantage of this elongated orbit is that, at some point, the earth is too close to the sun, and more energy is absorbed from the sun. Then, at another point, it is farther away, and this allows the earth to recover a bit from the high radiation that was absorbed during the high-intensity period.

The earth is like a chicken on a rotisserie. The center of the chicken is at a constant distance from the heat source. The chicken rotates and cooks uniformly, all over. Let us suppose that, instead of threading the chicken through the center from the tail end to the neck end, we threaded it through the left leg to the right breast. The rotation of the chicken would be lopsided. The right thigh and left breast would rotate closer to the heat source than the opposite counterparts. So, by the end of the cooking, the right leg and left breast would be overcooked while the left leg and right breast would be relatively undercooked.

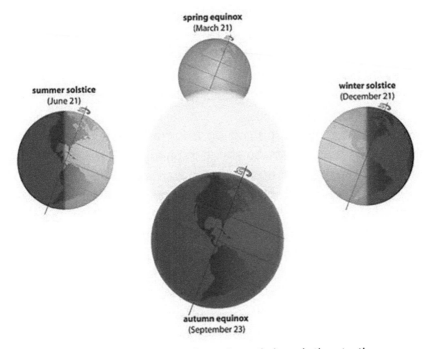

Figure 25. Relative location of earth in relation to the sun

The chicken-on-a-rotisserie effect is, in fact, what happens with the Tropics of Cancer and Capricorn. The short time when the sun's motion north to the Tropic of Cancer and south to the Tropic of Capricorn stops and the direction is reversed is enough to cause major climate effects in areas where the concentration of the sun's radiation is high.

In the Tropics of Cancer and Capricorn zones, the twenty-kilometer beam goes over the same area twice over a very short period of time (i.e., two to seven days). At the equator, for instance, the sun moves continually over the area in one direction. The sun does not return to the area until three months later. So, relatively speaking, the vegetation at the equator is less likely to go through the types of heat stresses experienced by the vegetation at the Tropics of Cancer and Capricorn.

Why Deserts Start on the Western Sides of the Continent

The general principle of air motion over land and sea has been described (see *Typical Motions of Water Vapor over Land and Water*). During the daytime, land warms up more quickly than water. Because of this, air from the sea or ocean moves toward land to occupy the vacuum left behind by rising warm air. During the night, the land cools more quickly than the seawater; at night, air moves from land to sea.

Assume this is happening on a large landmass such as a continent. The air moves as stated in principle. But we know the sun rises from the east and works its way west.

The African continent is a good illustration of this principle. Before the sun rises on the east coast, it will have been burning over the Indian Ocean for hours. Considerable evaporation of water will have happened before daylight on the continent. Air will be moving from the landmass toward the Indian Ocean.

When the sun rises over the land, it warms up quickly, becoming relatively warmer than the ocean. Humid air will move from the ocean toward land. The chances of rainfall are very good.

Over several hours, the sun crawls over the Indian Ocean. Daybreak reaches the Horn of Africa sooner than it does the western Sahara coast. As daybreak comes to the west coast, the land warms up very quickly, and air from the Atlantic Ocean sweeps toward land.

The problem is that the air over the Atlantic Ocean would have been cooling over a period of twelve hours. The capacity of the Atlantic air bringing water vapor over land is much lower than the capacity of the Indian Ocean, so the chances of precipitation on the west coast are lower.

What If the Twenty-Three-Degree Tilt Did Not Exist?

If the earth's axis formed a ninety-degree angle within the axis running through the center of the sun and the center of the earth, soon catastrophic things would happen. To begin with, there would be no seasons. Cold places would remain cold, and hot places would remain hot throughout the year.

The length of the days would be constant at any given point on the planet. Every day, the sun would rise precisely at a specific hour and set at a constant hour. The 1,000-kilometer radius beam of light from the sun would beat down onto the equatorial zone, producing virtually the same average temperatures at the equator day in and day out.

It is easy to see that the equator would have become a desert millions of years ago. The vegetation would not be able to survive the scorching temperatures on a daily basis. The land would dry up, making it impossible for lakes to exist for any period of time in the equatorial zone.

It is also plausible to speculate that global warming would happen quite quickly. The major convection currents that the rising heat would cause would lead to severe weather conditions.

The hot air would rise to very high altitudes and fan out north and south. Cool air would sweep toward the equator from the poles. This cycle would have caused the polar ice caps to melt much sooner. There would probably be no ice at the North or South Poles by now.

In the no-axis-tilt scenario, rainfall would probably only fall between the fifteen-degree to forty-degree latitudes. It would be at these latitudes that the warm equatorial air would mingle with the cool polar air to cause rain. It would be difficult to sustain conditions where snow could be created.

The twenty-three-degree-axis tilt makes our current conditions possible. However, the fact that deserts form along the Tropics of Cancer and Capricorn suggests the instability of our climactic condition and is probably unsustainable into perpetuity.

The presence of vast quantities of water on earth is both a saving grace and a curse. It is a saving grace in that there will be evaporation as long as the sun shines. The evaporation will most likely cause rainfall at some points on land. The rainfall will continue to sustain life on land.

The curse is that evaporation is a means of circulating solar heat on earth. The more water evaporation occurs, the warmer the planet will get. It is possible that a condition could be reached when rainfall in some places will evaporate before it hits the ground. This condition already exists in some parts of the earth today. It will most likely become more commonplace.

SUMMING UP THE EFFECTS

The observation of increasing carbon dioxide is an important clue in the trend toward global warming. Whether or not carbon dioxide is the primary factor responsible for global warming is still subject to debate. The mechanism by which carbon dioxide holds heat and the factors that trigger the subsequent release of heat should be explained more clearly. What has been confirmed is an existing correlation between a rise in temperature and an increase in carbon dioxide.

In view of the fact that carbon dioxide is one of the by-products of combustion, there is still a level of uncertainty as to what is really happening. Clearly, in addition to carbon dioxide, combustion also releases water vapor as a key by-product. What arguments have been advanced to discount water vapor as the primary cause of global warming?

It is known that water vapor is, in fact, transformed liquid water and is a gas by virtue of the fact that the liquid water absorbed heat and the heat changed its form. However, questions remain. Can we account for the fate of the heat of transformation when the vapor condenses back to liquid water? How much water vapor does the combination of combustion engines and working activities generate? Do we have an accurate accounting of the effects of the various water- evaporation activities?

How would global warming mitigation activities differ from those in which carbon dioxide is emphasized as the primary cause of global warming? How would the mitigation technology differ from that being proposed to reduce carbon emissions?

There are many unanswered questions regarding global warming. It is a subject with a wide scope and is possibly the top issue of this millennium. None of us were around millions of years ago when the Sahara Desert was formed. But it might be fair to conclude that the creation of the Sahara may have resulted in a global increase in temperature. Though there are some clues left behind, no one can authentically say what happened during the Ice Age and the subsequent melting of ice.

The best we can do is theorize and speculate on the cause (or causes) of global warming. The argument for greenhouse gases has been well presented. This argument puts considerable emphasis on carbon dioxide as the primary cause of the warming. And this argument leads to seeking solutions aimed at reducing carbon dioxide levels in the atmosphere.

Alternative arguments should be considered. The argument in this book is that the primary culprit is water vapor. To start with, combustion does release water vapor. We can conclude this by examining the molecular structure of common fuels. The simplest fuel molecule is that of methane. The chemical equation for this compound is CH_4.

When the methane molecule is burned, one molecule of carbon dioxide and two molecules of water are generated for every one molecule of methane. In short, more water is generated than carbon dioxide.

Scientific knowledge indicates that most combustion processes do not release much carbon dioxide. In fact, the immediate gas produced is carbon monoxide, which eventually gains additional oxygen after the fact but not during the combustion process.

The production of carbon monoxide during combustion suggests that the hydrogen part of the fuel molecule takes up most of the immediately available oxygen, leaving carbon in a bind. The consequence is that some of the carbon never gets oxygen and is precipitated as soot, which is pure carbon. The rest of the carbon grabs some oxygen but not enough, and the immediate by-product is carbon monoxide.

In this argument, the primary fuel in fossil fuels is hydrogen. The affinity between oxygen and hydrogen is so strong that no other element can compete with the reaction between the two gases. We can then conclude that carbon really is only a vehicle to allow us to transport hydrogen safely until combustion occurs. If the key fuel in fossil fuels is hydrogen, we must accept that the key by-product of combustion is water vapor.

The circumstance under which fuel combustion occurs is that heat is present. So, inevitably, the generated water vapor is invisible. However, since heat is a factor in the production of the combustion by-products, we must consider the transformation energy contained in the reaction. The water vapor carries the transformation energy. We do not feel the heat immediately due to the air-dilution effect. But the entire heat of a combustion engine is there in its entirety. By nature of the dilution effect, this heat could possibly travel far and wide over the globe before being dissipated into outer space or being absorbed by solid objects encountered by air.

The mountaintops, where snow cover is diminishing, are another symptom of increased water vapor in the atmosphere. There is very little or no CO_2 on mountaintops, but H_2O will get there if conditions are right. Obviously, more water vapor means heavier clouds and the increased likelihood of precipitation at lower elevations. So less and less water vapor gets to mountaintops, yet the mountaintops high winds cause snow-cover loss. Ultimately, mountains, particularly those in the tropical areas, will lose snow or ice cover.

The loss of snow cover on mountaintops has implications for the water supply at lower elevations. Rivers that now supply many communities with fresh water will dry up, further aggravating the global warming problem.

The Key Factors in Global Warming

Global warming is not caused by human activity alone, but it is fair to emphasize that we have exacerbated it. The two major factors in the phenomenon are the chicken-on-a-rotisserie effect and increased water evaporation, particularly due to agricultural activity, use of automobiles and other industrial and domestic activity.

The chicken-on-a-rotisserie effect has been explained above (see *The Chicken-on-a-Bad-Rotisserie Effect)*. The earth's tilted axis is helpful in slowing down what would have otherwise happened sooner. If there were not a tilted axis, the equatorial zone would have been turned into a desert thousands of years ago. The persistent radiation over the same zone, day in and day out, would have made it impossible for vegetation to grow along the equator, and the zone would have become the primary cause of global warming. Ultimately, desertification would spread north and south of the equator.

The tilted earth's axis shifted the focus of global warming for the Tropics of Cancer and Capricorn. These zones receive several days of consecutive solar radiation, much more than anywhere else on earth. While the consecutive days of repeated radiation are brief, the effect is clear. Most of the world's severest deserts occur in these zones. Also, the lack of vegetation cover in deserts makes them major factors in global warming. The air gets superheated, increasing the capacity to cause water evaporation over oceans.

Evaporated water is a major reservoir of latent and potential energy. The potential energy is evident in waterfalls, a major source of our hydroelectric power. But the evidence of the latent energy is there, and its influence on us has often escaped our realization.

Generally, people see clouds as cold and relatively calm. It is difficult to associate them with energy in the form of heat. The reality is that as long as water exists as a vapor, it contains the heat of transformation. The heat must be released before vapor turns to liquid. The heat exchange occurs when supercold air mingles with relatively warmer, moist air.

The transformation of liquid water to vapor is catastrophic; it is, in fact, explosive in an open atmosphere. The ratio of water to vapor volume can be as high as 1:1,600. In other words, one liter of water can potentially become 1,600 liters of water vapor in seconds over an open-tilled field.

GLOBAL WARMING MITIGATION STRATEGIES

In fact, *if* water vapor is the cause of global warming, then strategies focused on carbon sequestration are misdirected and will be ineffective. Worse still, strategies that promote the use of hydrogen fuel are downright dangerous. Hydrogen may sound like a clean fuel, but its primary and perhaps only by-product is water vapor.

Modern civilized society is built around burning fuel to improve life. In other words, modern society centers around exhuming sun energy that was buried billions of years ago and is now being released as new energy. This would work if the scale of the release of old energy was small. At the current time, the scale is so large that the effects are noticeable.

It is unlikely that modern society will reduce the need to recycle old archived sun energy. If anything, the demand will only get higher, because a large proportion of the world's population has not yet reached the level of comfort that is being enjoyed by the wealthier nations.

In spite of the gloomy prognosis, something can be done to forecast the effects more accurately. Some ideas are discussed below.

Monitoring Temperature

The world is in overdrive regarding monitoring climate change. Efforts range from measuring tree-growth rates in almost all countries on the globe to tracking ice-melt rates in polar regions. These are very helpful developments.

In addition to collecting real-time data, IPCC models old and new are forecasting what the future might look like. So far it has been demonstrated that real temperature changes are within the range of possible outcomes predicted by the IPCC models.

However, the most critical zones of the world (where most of the weather-forming events might be happening) are not receiving much attention. These are the Tropics of Cancer and Capricorn. These zones are in fact the hottest areas on earth, and their generated heat travels to all corners of the globe. An examination of the IPCC analysis shows that the temperature trends in the two hot zones are barely within the upper limits of the IPCC bounds of tolerance.

It would be prudent to establish remote accessed monitoring stations along the twenty-three-degree latitude (north and south). These monitoring stations would provide better data on future hot temperatures and the speed at which this will happen.

Generating Less Water Vapor

A close examination of activities that generate water vapor might produce surprising results. Innocuous activities such as irrigation, showering with hot water, cooking, and lawn watering might suddenly be found as not so innocuous. Also, many industrial activities involve boiling water. Food preparation and maintenance of hygiene all involve large-scale production of water vapor.

Among the worst offenders for generating water vapor are nuclear plants. These plants use water to cool the fusion materials as the nuclear reaction takes place. The hot water is then released into the environment or into oceans or lakes. Such releases will inevitably increase water evaporation rates.

The opportunities to release water vapor far exceed those of releasing carbon dioxide. For instance, boiling water on an electric stove will not release any carbon dioxide at all, but vapor is released.

Consideration should be given to assessing methods of reducing vapor production. From this perspective, what has seemed so far to be a straightforward problem becomes more complex.

Some suggestions for reducing water vapor production include the following.

Take baths instead of showering. Showering releases more water vapor per shower event than taking a bath in warm water.

Reduce the number of cooking events that involve water boiling; for instance, drink cold drinks rather than hot drinks.

Eliminate lawn watering altogether. Green lawns are great to look at, but we do not want the globe to fry for them.

Reduce broadcast irrigation in favor of drip irrigation. Broadcast irrigation results in more water being evaporated into the atmosphere than drip irrigation.

Convert steam and vapor generated by industrial production to liquid water at source. This will reduce the export of vapor to sensitive areas such as the Arctic, Antarctica, and glaciers.

Adaptive Technologies and New Strategies

The most rapidly growing means of generating water vapor are vehicles and other industrial machines that burn fossil fuel. The technology for controlling automobile emissions has evolved considerably. Today, cars emit fewer toxic gases than they did two decades ago.

Up to this point, the focus has been on removing carbon particles (soot) and various carbon-based chemical gases. The existing paradigm is that water vapor emitted by cars is clean and harmless.

The reality is that more than 60 percent of the air that comes out of a car tailpipe is water vapor. This is because fossil fuel originates from water, and when burned, the original water is reconstituted. Furthermore, it is reconstituted at a much higher temperature than it possessed at the time of fuel formation.

We tend to be fooled by the heat dilution that occurs immediately after the hot vapor comes. But the car heat is simply transferred to other water molecules in the air. However, as the number of vehicles increases, the effects will become more perceptible.

It should be possible to remove water vapor from car exhaust before it is released into the air. This requires a major paradigm shift and may take many years to achieve. To begin with, the public has been led to believe that water vapor is not an issue. Many governments are preparing to switch to hydrogen power. All the focus has been on reducing carbon dioxide emissions. There is so much momentum in that direction; additionally, major institutions are not designed to change direction quickly.

The redeeming feature is that carbon dioxide production correlates with water vapor production. In that sense, new strategies to address the reduction of water vapor production would merely involve minor modifications to the "reduce carbon dioxide" strategies.

A major difference would develop in the carbon-trading business. Instead of trading carbon credits, an index based on vapor generation assessment would be preferred. Hopefully, the carbon credit idea can be killed quietly without anyone being hurt. From the perspective of the countries lying along the Tropics of Cancer and Capricorn, the idea was a nonstarter. There is very little chance that those countries will grow trees in deserts in the near future. Their focus will be on maintaining the little arable land they have for agricultural use.

IS GLOBAL WARMING REALLY CAUSED BY CARBON DIOXIDE?

So we come back to the original question: Is climate change really caused by carbon dioxide? Opinions out there differ on this issue. The evidence exposed in this book seems to suggest that yes, carbon dioxide may be involved, but it is unlikely to be the main cause of climate change. In the opinion of the author of this book, water vapor is the primary cause because it carries with it latent energy that tends to persist in the atmosphere due to its nature of transfer from warmer vapor molecules to cooler vapor molecules rather than a dissipation as wavelength radiation.

When liquid water is transformed into vapor, the behavior of the molecule changes. The molecule behaves like an inflated balloon and may float to very high elevations—sometimes going as high as 10,000 meters above ground. This behavior of vapor is commonly observed when we see clouds. They seem to float so effortlessly and can be sighted at very high altitudes.

Furthermore, if one were to climb Mount Everest, he or she would surely encounter ice up there, and the only way it could get there is though floating up to high elevations.

Carbon dioxide, on the other hand, is a heavy gas. If you had a glass 1,000 meters high, and you poured one liter of carbon dioxide in it, all of it would fall to the lowest part of the glass. This illustration of how heavy carbon dioxide is is a common experiment taught in almost all high school chemistry classes. Oxygen is lighter than carbon dioxide, but for mountain climbers, it is well known that climbing a mountain higher than 5,000 meters can require the use of portable oxygen.

If humans relied on carbon dioxide for respiration, the upper limit for requiring portable carbon dioxide would probably be at 3,000 meters above sea level. The highest concentrations of carbon dioxide are at low elevation levels.

Given that we are experiencing accelerated melting of glaciers at high elevation, it is unlikely that rising levels of carbon dioxide is the cause up at those high elevations.

REFERENCES

Ångström, Knut. "Ueber die Bedeutung des Wasserdampfes und der Kohlensäure bei der Absorption der Erdatmosphäre." *Annalen der Physik* 308, no. 12 (1900): 720–732. English translation by Peter L. Ward with careful editing by Jochen Grocke, 2013.

Arrhenius, Svante. "On the Influence of Carbonic Acid in the Air upon the Temperature of the Ground," *Philosophical Magazine and Journal of Science*, Series 5, volume 41 (1896): 237–276.

Arrhenius, Svante. "The Effect of Carbon Dioxide (CO_2) and Water Vapor on Atmospheric Warming—A Century-old Question." Ofversigt d. Stockh Ak. 1901, No 1, p. 55 & 56 (1901).

Black, B. C., and G. J. Weisel. *Global Warming—A Historical Guide to Controversial Issues in America*. Greenwood, 2010.

Gain, A. K., H. Apel, and F. G. Renault. "Thresholds of hydrologic flow regime of a river and investigation of climate change impact—The case of the lower Brahmaputra River Basin." *Climate Change* 120, no. 1 (2013): 463–475.

Intergovernmental Panel on Climate Change (IPCC), 2007.

Nelson. "The Effect of Carbon Dioxide (CO_2) and Water Vapor on Atmospheric Warming." US Arctic Research Commission. Permafrost Task Force Report, 2003.

Lightning Source UK Ltd.
Milton Keynes UK
UKHW012015060521
383282UK00001B/61